The Field of Geography

General Editors: W. B. MORGAN
and J. C. PUGH

Agricultural Geography

In the same series

Political Geography

J. R. V. PRESCOTT

Agricultural Geography

W. B. MORGAN and R. J. C. MUNTON

ST. MARTIN'S PRESS · NEW YORK

Printed in Great Britain

For information, write:
St. Martin's Press, Inc.,
175 Fifth Avenue, New York, N.Y. 10010

Library of Congress Catalog Card Number: 72–76588

First published in the
United States of America
in 1972

AFFILIATED PUBLISHERS
Macmillan & Company, Limited,
London – also at Bombay, Calcutta,
Madras and Melbourne – The
Macmillan Company of Canada,
Limited, Toronto

Contents

The Field of Geography

Progress in modern geography has brought rapid changes in course work. At the same time the considerable increase in students at colleges and universities has brought a heavy and sometimes intolerable demand on library resources. The need for cheap textbooks introducing techniques, concepts, and principles in the many divisions of the subject is growing and is likely to continue to do so. Much post-school teaching is hierarchical, treating the subject at progressively more specialized levels. This series provides textbooks to serve the hierarchy and to provide therefore for a variety of needs. In consequence some of the books may appear to overlap, treating in part of similar principles or problems, but at different levels of generalization. However, it is not our intention to produce a series of exclusive works, the collection of which will provide the reader with a 'complete geography', but rather to serve the needs of today's geography students who mostly require some common general basis together with a selection of specialized studies.

Between the 'old' and the 'new' geographies there is no clear division. There is instead a wide spectrum of ideas and opinions concerning the development of teaching in geography. We hope to show something of that spectrum in the series, but necessarily its existence must create differences of treatment as between authors. There is no general series view or theme. Each book is the product of its author's opinions and must stand on its own merits.

W. B. MORGAN

J. C. PUGH

University of London,
King's College
August 1971

Maps and Diagrams

Tables

Acknowledgments

The authors wish to thank the following:

The Secretary-Treasurer of the *American Journal of Agricultural Economics* for figure 3.1

The Editor of the *Journal of Agricultural Economics* for figures 5.1 and 5.2

The Director of the Agricultural Economics Unit, Cambridge for figure 6.1

Pergamon Press Ltd for figure 7.2

The editor of *Economic Geography* for figure 7.4

Figure 7.5 from *The Location of Economic Activity* by E. M. Hoover 1948. Used with permission of McGraw-Hill Book Company.

The Macmillan Company, New York for figure 7.6

The Geographical Association for figure 8.6 which appeared in *Geography.*

The Controller of Her Majesty's Stationery Office for figure 9.1 which is Crown Copyright.

The Hon. Editor of the *Transactions of the Institute of British Geographers* for figures 10.1, 10.2

Introduction

Interest in the spatial patterns of agricultural activities provided one of the earliest and most firmly established foci of study in human geography. Moreover, it was on agriculture that von Thünen, the founding father of economic location theory, concentrated his attention in the 1820s (von Thünen 1826), while one of the first attempts to introduce non-profit maximizing assumptions into location theory was again concerned with agriculture (Brinkmann 1922). Despite these early developments in related fields of study, the acknowledged importance of agricultural studies by economic geographers, the emphasis of certain historical and cultural geographers on the evolution of landscapes and diffusion of domesticated plants and animals (see Sauer 1952, for example), and outstanding work by such pioneers as O. E. Baker (Baker 1926–32), progress in the development of agricultural location theory has been slow. Agricultural geographers, before the 1950s, had been almost obsessively preoccupied with explanations for distribution patterns derived from study of the physical environment alone. Related studies in economics or the relevance of economic principles were largely ignored. This one-sided approach, often severely limited in its understanding of how the factors of soil, climate, and slope related to agricultural activities, was frequently linked with gross generalizations about farming types, which examined individual farms only as case studies, or with descriptive land-use which, on occasions, ignored the existence of farm units altogether.

To some extent a single discipline of agricultural geography is difficult to conceive. It could be argued that agricultural location theory is either being developed or could be better developed in other fields, and, in support of this, one may point to the fundamental contributions to theory of economists such as von Thünen and Lösch. This view stems in part from our difficulty in understanding the agricultural patterns involved. Such understanding must depend on examining not just the pattern but also the nature of the relationships obtaining between the phenomena concerned. In agricultural geography these relationships are of many different kinds, so different that they cannot be incorporated as relationships into a single system of laws, even though they may be described as spatial phenomena by a single if complex geometric pattern. Agricultural investigation is concerned with economic, cultural, and biological relationships. Although we may measure some of the biological and cultural relationships in financial terms, we may never entirely put all the variables and constraints present in farming on to a monetary scale, nor may we appreciate the choice of crop or livestock confronting the farmer unless we learn something of farming as a way of life and something of the farmer as a member of a rural community. This problem of reconciling different sets of variables has been described more generally by Emery and Trist, '. . . the laws connecting parts of the environment to each other are often incommensurate with those connecting parts of the organization to each other, or even those that govern the exchanges' (1965). Thus, for example, the laws connecting the actions of the javelin thrower in sighting and throwing his weapon cannot be used to describe the course of the javelin, which is affected by variables linked by meterological and other systems (Barker and Wright 1949). If, however, we leave the development of location theory to disciplines such as economics, sociology, or botany, we shall not gain a full understanding of the location of crops and livestock as relatively few economists, despite the work of von Thünen and his successors, and even fewer sociologists and biologists have been interested in location problems. Thus, although we may have difficulty or even find it impossible to reconcile all the relationships between agricultural phenomena into a single system of laws, there is a real need to encompass them in a single system of learning.

1 Nature of agricultural geography

A scientific agricultural geography is not concerned simply with describing the nature of farming in particular places. Although such description must always have some interest for the agricultural geographer, his main objective must be to understand the spatial aspects of farm enterprises, that is crops and livestock, whether considered individually or in groups, and of farm operations. The transport and marketing of agricultural commodities must also be of major interest to him, although such studies have been largely neglected in favour of production-oriented investigations. However, this does not exclude the possibility that in the future agricultural geographers will give a higher priority to these related features.

Agricultural geography is a part of economic geography (for discussion, see Coppock 1968). It is concerned at present essentially with production processes, and in the first instance must seek to classify and comprehend them. It must therefore employ the relevant basic concepts and principles of economics, for agriculture is an economic activity, no matter how important knowledge of soil and moisture relationships may be to a farmer. As economics may be said to be a science of choice, so agricultural geography is concerned with choice, but with a special locational aspect. Thus although the location of any enterprise may be seen generally in relation to known economic and biological factors, it is clearly the product of decisions made by many farmers either to develop or not to develop that enterprise, or of their ignorance of the ways in which it might fit into their scheme of operations or farm systems. Since the farm is the decision-making unit it should be a major focus of interest of the agricultural geographer, both in research and in the development of theory. This is not, however, to deny the importance of work at other levels of study.

The scale problem is fundamental in geographical studies (Haggett 1965, 263–5; Harvey 1968). In agricultural geography data are collected and generalizations made most frequently at four major levels:

1. The nation
2. The agricultural region, however defined

3. The farm
4. The field.

At the national level there is frequently an abundance of published statistical information, often provided annually and referring to administrative subdivisions, making possible some regional comparison and inference.

Agricultural regions may be larger than some nations, but are normally subnational units defined by whatever criteria are available or needed for a particular kind of study. These criteria may be generic and refer to particular enterprises, groups of enterprises, type of farm practice or structure, or may refer to functional relations such as obtain, for example, in the regional orbit of a market or of a factory treating agricultural produce. Such regions, however defined, are frequently poorly provided with data since they rarely fit the framework of administrative regions used either for data collection or for data publication.

The farm is largely self-defining in area, although its classification as a farm-type may be highly subjective and, where business arrangements obtain between otherwise separate holdings, there may sometimes be difficulty in recognizing the spatial limits of the decision-making unit, or farm firm. Geographic interest concentrates on the location of the farm in relation to markets, the organization of the farm as a spatial system of operations and the farm site, especially land quality. Farm data must frequently be collected directly from the farmer by the research worker, for normally farm data collected for national records are only available for study in aggregate form.

The field is the basic unit of agricultural land-use. It can be viewed in a static sense with reference to a given moment of observation or a given cultivation season, dynamically as a part of a field system cultivated in a cycle of operations, or historically as the current end-product of a number of operation cycles some of which may never have been completed before a change was required in the system.

Empiricism and deduction

Traditionally agricultural geography has been concerned with describing, classifying and explaining that which exists in areas. Such studies depended mainly on inductive reasoning using empirical methods in which generalizations were made from features observed. Deductive methods by which hypotheses were first developed, and then tested by examination of the facts, have only recently played an important role. Such methods have rarely been employed, however, to explain that which exists in areas, but more usually to predict that which might have existed had certain assumptions been satisfied. Generally the little deductive theory that has been developed in agricultural geography has been concerned with seeking optimum distributions, usually the distributions which will give most profit assuming that all farmers are possessed of equal reasoning powers, are equally able to manage the possible enterprises and are equally supplied with information. Analysis of the relationships involved may be presented in the form of optimizer models (Haggett 1965; Pred 1967; Chorley and Haggett 1967). Sub-optimal behaviour on the part of farmers to achieve some satisfactory return, not necessarily the best, has been examined by satisficer models (Harvey 1966). Others have argued the case for sub-optimal analyses based on such techniques as simulation, since input data, production, and price variability and management constraints seldom permit one to know the maximum profit system with any certainty, or to desire its adoption as the best solution for current problems (Donaldson and Webster 1968). If the observed world is required, satisficer models may be best, but scientific objectives depend on the problem to be solved. For Lösch, for example, the question of the best location was far more dignified than determination of the actual one (Lösch 1954, 4) and optimal rather than actual patterns of production have been the concern of several agricultural economists using linear programming methods (Heady and Egbert 1964; Henderson 1957; Whittlesey 1967; Shaw 1970). Such knowledge may be of considerable theoretical importance, even though it lacks immediate practical application, and its results may be difficult or impossible to test because of the 'uniform plane' problem, that is the lack of suitable farms where all factors except those to be tested can be observed to be constant, or controlled in a series of experiments. Even satisficer models leave a considerable gap in attempts to explain the observed world and attempts to close this gap, apart from those which employ mathematical theories of chance, look more and more to empirical methods based on statistical techniques such as factor analysis (Kendall 1939; Henshall 1967).

Prediction, the understanding of what should exist for any given moment in time, must also be able to cope with the problems of change from one agricultural system to another, different rates of diffusion of agricultural techniques and problems of individual readiness to accept change. It must take account of differences between farmers, not only in their management efficiency, but also in their attitudes to farming. Of considerable importance here is the increase in Europe and North America of part-time farmers with other interests, and of hobby farmers whose interest in the land may in only small part be concerned with financial profit. Ideally agricultural geographers would like to be able to predict the distributions of enterprises and agricultural practices which would obtain were one or more of the factors affecting their locations changed. Such work could be of enormous economic importance. However, among the many problems involved is that of the unpredictable nature of the market in agricultural produce, making the costs of operating the models involved and of collecting the data required enormous. We have at present to be content with much less, and vital though the development of deductive work is for the future of the subject, empirical work and comparison of observations can still yield considerable information and important generalizations. For example, the study of 'best holdings' rather than the observation of all has been justified on the grounds that the practices of the best agricultural holdings observable today will be typical of most farms in a few years time. This assumption is implicit in many agricultural innovation diffusion models (Hägerstrand 1952; Bowden 1965).

Economic concepts and principles

The fundamental concepts of economics are fundamental also to agricultural geography:

(i) the USE of resources of time, energy, property, goods, techniques, money, information and sources;
(ii) the CHOICE of alternative enterprises, farming systems, methods of production, transport and markets;
(iii) the EXCHANGE of goods or of money and goods, which plays an important role, even in so-called subsistence economies;
(iv) the SCARCITY of resources making it necessary for individual farmers to make the best use of those they have.

Production, or the SUPPLY of agricultural commodities, depends on the satisfaction of DEMAND, expressed frequently in the market. SUPPLY and DEMAND govern PRICES, and in turn are affected by them. Prices reflect our estimates of the VALUE of things in terms of exchange (the price we think we *should* obtain) or use (that is their utility in satisfying needs), the SCARCITY of things and their costs of production and exchange. We may analyse production processes by the examination of OUTPUT or the results of production as agricultural goods together with waste products, and INPUT or the factors of production, consisting of land, labour, financial resources, buildings, equipment, power, feed, seed, fertilizers, pesticides, lime and other supplies.

The basic principles governing the location of production, where the variables are entirely economic, are derived from the law of SUPPLY AND DEMAND. Where supply is constant, the higher the demand, the higher the price, and where demand is constant, the larger the supply, the lower the price. The higher the price, the less buyers are ready to purchase and the less sellers are ready to sell. In a theoretically perfect market the activities of buyers and sellers should result in an oscillation of prices around an equilibrium price, i.e. the price at which the supply and demand curves intersect (fig. 1.1) and at which there will be no unsatisfied buyers and no unsatisfied sellers. A change in demand or in supply should lead to the displacement of the appropriate curve and the movement of the point of intersection or equilibrium to a new level. In location terms we may imagine a given market exclusively served, for the purposes of simplification, by a sufficient number of farms to

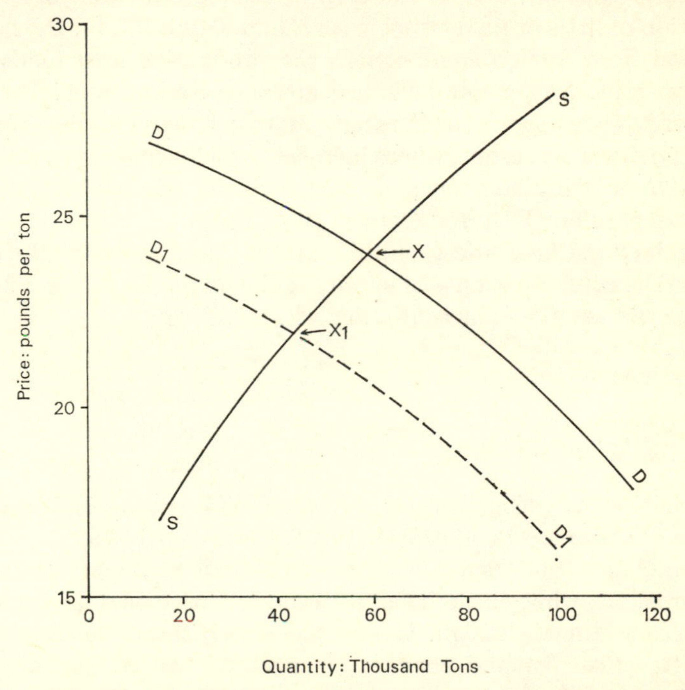

1.1 *Hypothetical supply and demand curves in a perfect market.*

SS *Supply curve showing amount sellers are ready to sell at given prices*
DD *Demand curve showing amount buyers are ready to buy at given prices*
X *Point of equilibrium price*
D_1D_1 *Demand curve resulting from reduction in demand and showing amount buyers are now ready to buy at a given price*
X_1 *New point of equilibrium price for reduced demand*

satisfy demand. Should demand increase the production area may expand to include other farms. Should demand decrease the production area may become smaller. Here we have to make additional assumptions that increased demand cannot be satisfied better by more intensive production on those farms meeting existing demand or that decreased demand will not result in a general decrease in production on all the farms concerned. Similarly where demand is constant but a new technology makes possible higher yields per acre, supply must still remain constant if prices are not to fall. In consequence, other things being equal, the area under production must decline. All over the world we can see the expansion of agricultural production in the face of increasing population, rising living standards in some countries, and the increasing demands for raw materials made by manufacturing industry. In the more developed countries total food consumption per head tends to be fairly constant and in the case of food production for a steady population of little or no increase with a high degree of, for the farmers, insulation from foreign competition, the production area tends to remain constant. Any major technical advance leading to higher yields may therefore decrease it, as in recent years in the production of certain British horticultural crops, where increasing yields may have countered the effects of rising demand and even reversed the tendency towards acreage expansion (Best and Gasson 1966, 56–8).

To understand how prices, supply, and demand are related we must examine the rate of response of supply and demand to changes in price. This rate of response is measured by ELASTICITY:

$$\text{Elasticity of demand} = \frac{\text{Percentage change in quantity bought}}{\text{Percentage change in price}}$$

$$\text{Elasticity of supply} = \frac{\text{Percentage change in quantity offered}}{\text{Percentage change in price}}$$

Elasticities vary considerably between different agricultural commodities. Thus in most countries with a high standard of living elasticity of demand for staples such as wheat flour is low, consumption being fairly constant, while elasticity of demand for foods such as fruit, salad vegetables and meat is high. The income elasticities of demand, i.e. the extent to which demand changes as income changes, for such commodities are high. In consequence, producers of such crops look to markets in countries of high living standards. Producers of wheat, although they also sell mainly to markets in such countries, can only expand production and sell more if they can find growing markets in countries of lower standards of living or in those willing to substitute wheat for another staple. In consequence, we have the spectacle of wheat dealers seeking markets in Africa and Asia, both in relatively poor countries and in richer countries where they hope to substitute wheat for rice. If we examine supply the elasticity of tree crops is clearly less than that of annual crops. Changes in world demand for groundnuts can meet with a response in the next planting season, but increasing world demand for cocoa or coffee must wait several years before fresh plantings become effective. Declining world demand for cocoa or coffee can be offset to some extent by withholding supplies, which in turn may be released when demand increases. There are, however, limits to storage and to other means of offsetting the problems of fluctuating prices, and generally tree-crop producers find much greater difficulty than other farmers in coping with market problems. An important factor here is producer SUBSTITUTION or consumer CHOICE of one agricultural product instead of another as prices change, e.g. one grain for another, alternative fruits or lamb instead of beef. Synthetic substitutes such as artificial rubber and fibres such as nylon have very much affected the markets for agricultural products, although they may also be said to have created markets of their own.

The principle of COMPARATIVE ADVANTAGE is important in location study. At this beginning stage we may consider two aspects:

(i) that a product *tends* to be produced by those farmers whose ratio of advantage in the production of it is higher;
(ii) that the choice of crop or livestock enterprises or combination of enterprises *tends* to be determined by comparative advantage in terms of profit or net return.

In the latter case we consider comparative advantage within the farm unit and the theoretical tendency to select those enterprises which will give the highest income. Thus it may be better to produce an indifferent crop of potatoes than an excellent crop of wheat, depending on the differences in demand and costs for potatoes and wheat (Black 1953). Comparison of yields and areas under production in an agricultural atlas frequently shows that areas of highest production are not necessarily areas of highest yield (Coppock, 1964*a*).

Another important law is that of DIMINISHING RETURNS. The law or principle of diminishing returns states that *successive applications of labour and capital to a given area of land must ultimately, other things remaining the same, yield a less than proportionate increase in produce* (Cairncross 1966, 58; fig. 1.2). The law may also be applied to land in that land varies considerably in quality and therefore extensions of cultivation may take up successively areas of poorer land with, in consequence, ever poorer returns if inputs remain the same. We can conceive therefore of decreasing returns for each additional unit of input or each additional unit of even poorer farmland. The point at which the farmer decides it is no longer worth while to continue increasing his input or adding hectares is the MARGIN OF PRODUCTION and the last unit of input is the MARGINAL UNIT. This may be a perceived rather than actual margin and as an economic notion should not be confused with MARGINAL LAND which is frequently used to mean land of low productivity in absolute terms, that is physically or ecologically marginal. For example, potatoes fail if grown in soils with a pH of less than 4·0. Land with a pH of 4·0 is therefore physically marginal for potato production and this margin can be spatially located.

Similarly, MARGINAL LAND should not be confused with the concept of the MARGINAL FARM. A marginal farm has been defined as a unit of production that cannot regularly yield sufficient income to provide the farmer with earnings at least comparable with those of a farm worker, as well as providing interest at the going rate on his capital investment, after business and housing allowances have been made (Ellison 1953). As such it would be wrong to assume that marginal farms will inevitably be on physically marginal land, although there is a marked correspondence between their distributions. A marginal farm may be on excellent land and be very small or badly managed (Chisholm 1966, 46–56; Symons 1967, 80–4).

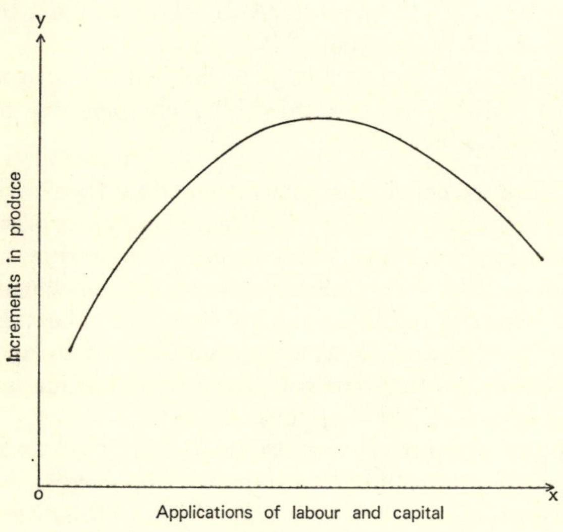

1.2 *Curve of productivity to illustrate the law of diminishing returns.*

Diminishing returns are associated not only with lower yields but also rising costs. On poor land, inputs have proportionately to be much greater than on good land to achieve comparable outputs, and for the most part the more profitable solution is to reduce inputs, even though this means lower returns, and to raise income by operating larger units of production. Hence the phenomenon of extensive grain production and ranching on low-quality land. The notion of a high-cost, low-productivity, or low-profitability limit to cultivation may be regarded as the EXTENSIVE MARGIN, which will fluctuate spatially with fluctuations in market prices. We may also think of another limit where improving conditions of production and changing costs make it more profitable to develop another enterprise. This may be termed the INTENSIVE MARGIN (Chisholm 1966, 51–3). The extensive margin of one enterprise may be the intensive margin of another and where the two coincide they are described as the MARGIN OF TRANSFERENCE. Such margins may be considered conceptually as separating zones each dominated by a particular enterprise or group of enterprises and farming may be said to be marginal for both systems of production at the contact point. Farming at the margin does not have to suffer from low returns as both systems may realize substantial profits. Its main characteristic in an uncertain economic climate should be frequent changes in location as the market prices of farm products fluctuate in relation to one another. In actual practice such changes will be limited by many other considerations, particularly of capital and of the technical experience of the farmers concerned. Actual land-use zonation does not have to occur for the concept to be valid and useful. Zonation will not occur, for example, in an area of highly differentiated land quality or managerial ability, but it is not so much the phenomenon of zonation that is of importance, as the relationships between the systems of production involved.

These notions of margin and of diminishing returns have provided the bases of a number of important economic theories of location with respect to the market, involving differences in transport costs, which we shall examine in chapter 7. Theories which deal with differences in land quality have as yet only limited application in the general body of agricultural location theory, perhaps surprisingly in view of the obvious importance of land quality as a location factor in any agricultural study and the considerable attention devoted to 'physical factors' by agricultural geographers. The earliest of these theories was that of ECONOMIC RENT (Ricardo 1817) or rent paid for the use of land alone. It differed from actual or contract rent which included payment for buildings and improvements. Economic rent could only arise from the differences between lands and was a consequence of the differences in costs of production and not a cause. Also it did not affect prices but was governed by them. Theoretically, lands at the very limit of production paid no rent. Rising prices will raise rents and push margins on to poorer land. However, on the margins of transference rent does enter into the costs of production since the differing rents even theoretically demanded by different enterprises become themselves a factor in choice. Land can be rented for one enterprise only if its use is denied to others, and the rent which it may yield under the most profitable of these other enterprises does enter into the cost and price of the enterprise chosen (Cairncross 1966, 294–5; Clark and Haswell 1964, 94). The notion that rent depends on marginal and not average production we owe initially to Ricardo. It was developed by Jevons (1871) and Marshall (1890).[1] Theoretically, the marginal return should be the same on

1. Marshall claimed he borrowed the concept of the margin from von Thünen (Marshall 1890; Hall 1966).

all lands, although the total return to labour on the more fertile land may be greater than on a similar area of land near the extensive margin, 'for the more fertile land is cultivated more intensively until its marginal return does decline to that obtained on the least fertile land in cultivation' (Cohen 1949, 27). But it is the marginal return which sets the limit to profitability and therefore to economic rent, and therefore it is not a question of a simple difference in fertility but of the rate of decline of successive units of output for the application of additional units of input. This notion of economic rent is useful in part in understanding the operation of production factors. In addition we must assess the effects of farm size or of additional input factors such as labour and management which today can more easily find alternative employment than they could in Ricardo's time. A useful technique is to estimate marginal productivity by means of a production function designed to examine productivity in terms of varying inputs of labour and capital (Clark and Haswell 1964, 98; Clark 1969). The Cobb–Douglas function, originally designed to examine quanta of industrial production in terms of varying inputs of labour and capital, is generally used in the form $P = L^a A^b$, where P is production, L the input of labour, A the area of land to be used, a the marginal productivity of labour, and b of land as a proportion of total product.

Cultural constraints

There is no general theory of social relations which as yet we may apply to agricultural location. However, social differences within communities and differences between communities affect scale and type of farming operations, together with choice of enterprise. Farming is the world's most widespread activity and in many countries its importance dwarfs that of all other activities. It is hardly surprising therefore in view of world-wide dependence on farming, that its operations should be the subject of social constraints and taboos, or that land should be set apart from other forms of property, be vested in the community, or even be regarded as an object of religious worship. We tend to compare the individualism of the commercially minded farmer of Western Europe or North America, subject mainly to the demands of the market, with the sense of family obligation of some African or Asian farmers, members of a large extended family, producing food mainly to satisfy that family's immediate needs. Such comparisons are frequently overgeneralized and unsound. A subsistence economy is still an economy and a market still operates even if it consists for the most part of the demands of the farmer's family. Nevertheless, to obtain the basic output required, peasant farming families are frequently willing to raise inputs of labour to remarkably high levels, especially in overcrowded areas where land is short. Although the extended family provides a form of social insurance for its members its obligations minimize the inducement for economic improvement and discourage innovation. The social system sets its imprint on holding and field systems, and on the settlement pattern with its related problems of accessibility to fields. Some peoples live in large compact settlements of several thousand population. The farthest fields may be more than ten miles away and distance to fields may become a problem, even a major factor in production.

Throughout the world farming is for the most part a family business. The bulk of agricultural production comes from small units worked entirely, or almost entirely, by family labour. Even in Western Europe and North America where advanced commercial systems have been developed family farming is still widespread and in some countries even dominant. In manufacturing industry the family business in certain trades still plays an important role, but for the most part output of manufactured goods is dominated by public companies and corporations. Despite the encouragement given in industrial countries to the development of larger agricultural businesses there is a sense in which family farming is increasing as a result of labour leaving the land, the greater use of machinery and the more efficient use of family labour. This does not mean to say that the outlook of farmers will become once more like that of their peasant ancestors, but it may mean that decision making with regard to the operation of the farm business will be more affected by family and social considerations than would have been the case had a permanent labour force been involved.

The spatial density of farmers depends on the type, size and structure of farm businesses. Market gardeners on smallholdings are normally close together whereas specialist grain farmers may be widely separated. Even where village communities exist and provide social nuclei for agricultural workers, rural densities of population are frequently too low for the provision of social services and amenities on the scale, of the standard, and at the price of those to be found in towns. Conditions of poor social provision are unattractive where agricultural operations have to be at their most extensive to succeed. Such areas tend to suffer from chronic depopulation, labour shortage and ageing populations. They may seek to develop tourism as a substitute activity for farming, but finance for social improvement is often limited as the costs of improvement are greater than in the better rural areas or the towns (Wibberley 1954). In the better farming areas in developed countries considerable social-status divisions have developed within the farming hierarchy between the high-status farmer, with usually a large farm and social contacts beyond the local community, and the low-status farmer, with a small farm, who is tradition oriented and does not identify with the social groups in the village with whom he has economic parity (Pahl 1965; Thorns 1968). These differences are reflected in major differences in crop choice and in the rate of acceptance and application of new techniques.

Physical and biological environment

In the past, general physical or biological explanations for agricultural location were accepted, partly because studies were made mainly at a macro-level where crop distributions tended to fit physical or biological distributions, and partly because geographers assumed that physical geography was itself basic to all human geography. This attitude still persists even in the chief textbook on the subject (Symons 1967) and in almost all descriptive economic geography texts (e.g. van Royen and Bengston 1964). Certainly the physical and biological factors are among the most important to be considered. Small temperature differences, for example, can effect enormous savings in production costs, even where techniques radically transforming the environment are employed, as in glasshouse production (Bennett 1963). Transport costs have been proportionately reduced (Chisholm 1962, 183–90), markets and distribution of goods have become much more effective, and in consequence as commercial agriculture has developed so the influence of other location factors including the physical and biological has apparently increased, despite the so-called ever-growing mastery over environment. Moreover, increased intensity in the use of soils frequently means increased sensitivity to environmental differences as corresponding proportionate differences in the operation of a system of higher inputs and higher outputs usually involve larger cash expenditures and higher gross incomes, although not necessarily greater profits. These closer relationships are the product of improved economic, chemical and engineering techniques combined under more knowledgeable and more effective management. Moreover, despite the apparent tendency of crops to occupy the most suitable areas for production in ecological terms, nevertheless at a time of reduction in producing area the worst portions are often surrendered no faster than the better lands. Chisholm concluded that 'adjustments in area may as probably occur on the good as on the poor wheat land, with the consequence that in more recent times no apparent correlation exists between crop-area changes and yields in the United Kingdom' (Chisholm 1966, 54–5; Moore 1960). There is some doubt about this conclusion, however, as the correlation between yields and land quality at any one time may be poor. Nevertheless for the purpose of argument we may construct a simple theoretical model showing an area in which one particular environmental factor has a dominant effect on crop distribution, in which there is perfect competition in the market, in which farmers' knowledge and decision making are assumed to be perfect, including correct estimation of future market prices, and in which transport costs have no effect, then the resulting pattern of a given crop and factor isopleths, e.g. isohyets or isopleths of insolation or frost occurrence, might appear as in fig. 1.3. The area shown under the crop would be sufficient to satisfy the market demand and the nearest factor isopleth to the areal limit of the crop at A–A, could be regarded as being a significant limiting factor. If, however, the demands of the market should increase, or some government should improve the strength of its price support, we may imagine,

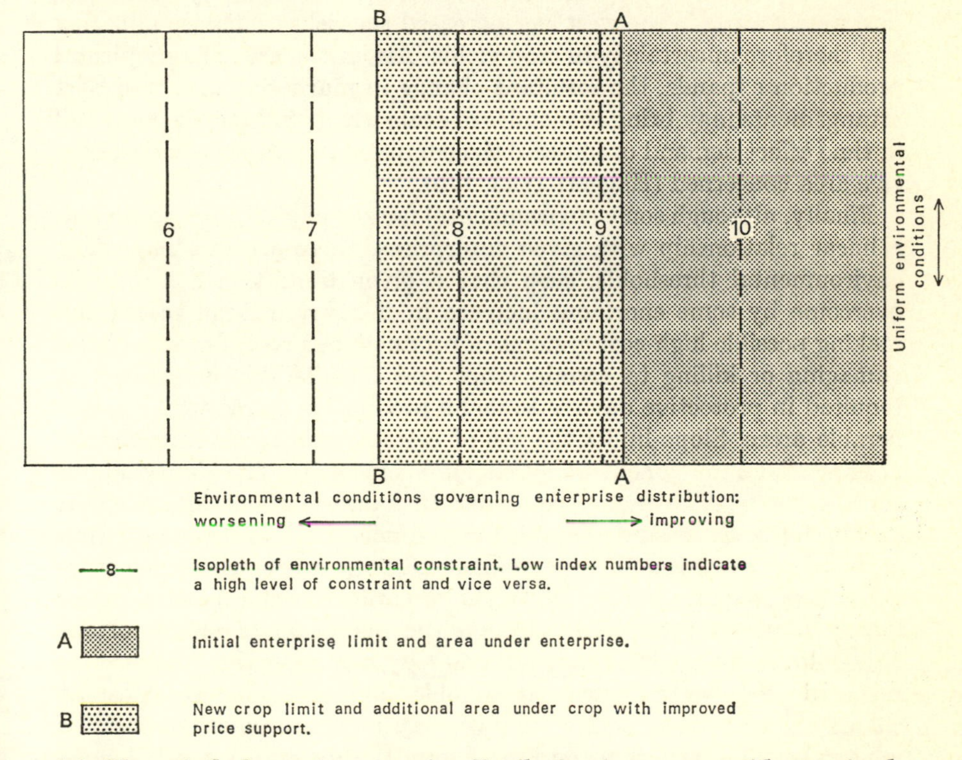

1.3 *Theoretical change in enterprise distribution in an area with one simple environmental variable and variation in price support, assuming perfect conditions.*

under perfect conditions, an expansion in the area of the crop and a new limit at B–B would be established, again showing excellent correlation with the environmental factor but coincident with a different value isopleth. Simple areal research of the static kind, common enough in elementary geographical work would fail to indicate the complexity of the relationship even at this simple level. In this case the environmental relationship determines the shape of the crop area, while the economic relationship determines its extent. In models assuming constant environmental factors this is inevitably the case. Climatic conditions change, however, soils may be modified and drainage improved, so that changing environmental factors may also influence areal extent. Again we have tended to assume in our model simple linear relationships, but this is not always the case. A change in quantity of a given environmental factor, for example, may result in changes in other interrelated factors. For example, wheat and barley distributions are related to rainfall quantity and incidence, but so are the distributions of certain diseases which affect wheat and barley. The spread of barley in Britain into wetter areas in the west has increased the risks of disease in barley and therefore uncertainty in crop yields. Also if the area of agricultural production expands, the resultant change in marketing and transport arrangements may bring economies of scale which will reduce costs and in time affect demand or increase the competitive power of the productive area concerned (Haggett 1965, 175).

Finally, although some environmental factors exhibit in certain areas a linear relationship with given enterprises, there are also important environmental thresholds, such that a given limitation can only be overcome by some enormous increase in demand, raising prices and making possible high-cost systems of production. Frost, for example, is damaging or killing to certain crops and considerable investment is required in protective devices to make production worth while in vulnerable areas. Slope limits the use of certain mechanical implements, needing the employment of much more expensive machinery above certain angles. The use of combines to harvest wheat limits wheat production to those lands suitable for combines, i.e. less than approximately 10° slope, without large stones and with a soil whose structure will not deteriorate under the weight. Combines thus become a major factor in wheat production through the operation of an important threshold. A change in combine design may well alter the relationships involved for technological thresholds are subject to constant change. Thus a technological change may reduce production costs or remove certain limitations on production, leading to a spread of some enterprise distribution. Hence the spectacular advance of new varieties of wheat and of hybrid maize.

2 Enterprises and systems

Agriculture is the purposeful tending of crops and livestock (McCarty and Lindberg 1966, 204). Apart from intensive livestock raising or crop production by hydroponics it involves management of the soil by tillage or by chemical treatment, by the establishment of a permanent grass sward or by the management of self-sown grasses and herbs. Whatever the technique employed the production process is essentially biological, differing fundamentally from most manufacturing processes. The production process is geared to the life cycles of either animals or plants or both in combination in a farm system. Much of the production process is outside the control of the farmer. He may assist or impede it, but in most cases he is uncertain of the results despite considerable advances in agricultural science. In any case more refined techniques offering greater measures of control may raise costs and be regarded as uneconomic. The time element is of major importance in the production process and in several ways:

1. *Seasonal rhythms* of enterprise life cycles which govern the differences in timing not only of labour demands but also of use of equipment, application of seeds and fertilizers, provision of feed and operations such as grazing or tillage. Some seasonal rhythms are fairly regular, e.g. sowing and harvesting, others, especially those dependent on weather, may be highly irregular. Seasonal patterns of production affect market prices. Attempts to produce out of season generally aim at higher prices, but at the cost of greater inputs. Enterprises may have different peaks of labour demand during the year and if the farmer can select enterprises with different peaks he may, if the demands are not particularly specialized, even out his pattern of labour-use. If not he will have sharp rises and falls in labour requirements.
2. *Non-seasonal rhythms* consist of such activities as twice daily milking and various beef-producing periods including rearing calves for sale as stores which are then fattened and sold at 12–24 months old, barley beef in 10–12 months or autumn calves fattened on summer grass in 15–18 months. In Britain milk yields a regular return paid by a monthly cheque whereas most crops provide only an annual payment.
3. *The cycling of production* occurs where there is a regular repeated pattern of rise and fall in production over several years. Production

cycles may be observed in a number of enterprises and mean that careful judgement must be required when examining productivity in any one year or any number of years, or in comparing one or more years. The problem has been stated by Harvey (1968) and demonstrated in a study of hop-growing in Kent (Harvey, 1963) where care is taken to compare data only for years corresponding to the peak and trough dates in the cyclical fluctuations. Considerable study has been made by economists of cyclic production in pigs, cattle, and crops (e.g. Gruber and Heady 1968; McClements 1969). Enterprise production involving a rotation round the farm and of uses of each field may be regarded as cyclic, but normally aims at a steady level of inputs and productivity.

4. *Periodic change* in seed, in best choice of livestock breeds, in techniques of production, in building standards and requirements, and in market prices comes with increasing frequency in this current period of rapid agricultural innovation. Such change can be rapid enough where rotations are in operation to cut a cycle of activity before it is completed. Whereas in the past long periods of relatively stable systems of production have distinguished agriculture, and farmers have been regarded as generally conservative in outlook, today there is constant change in pursuit of improvement, that is raising output and reducing input or using existing inputs more effectively. In Europe and North America the need now is to find suitable temporary buildings to accommodate short-lived systems.

Agricultural production is the most extensive form of production in its use of land other than hunting, fishing, gathering and forestry. As it occupies a great deal of space, the amount and quality of land available are vital production factors. Farmers in general cannot afford to pay as high a price for land as home buyers, manufacturers or people engaged in service industries. The demands of the latter can force prices to levels uneconomic for farmers. When home buyers wish to buy farms for the sake of rural residence farmland may be bought at a price too high for a profitable return from farming. Probably most of the world's farmers do not own their land but are either tenants, farm for some kind of share of the crop, or hold only the use right to their land, the full rights in which are vested in a lineage group or community. Lack of permanent right in land brings its own problems of crop choice and therefore of location. Tree crops, for example, are usually avoided on rented or short lease land or on land shared with other members of a lineage. Where improvements will be lost to the landlord or to a new occupier the tenant usually avoids making them. Leasing agreements may lay down specific conditions with regard to trees, use of soils and expiry of lease in relation to harvesting.

Much commercial agriculture tends to be distant from its main markets, and its produce is mostly bulky, although in some cases, e.g. wheat, modern handling and transport methods have reduced transport costs. Although farmers may sell in many markets and some do, in practice most farmers sell in very few markets, often only one, or through an agent or to a dealer. Early spatial models of the farm–market relationship mostly depicted farms or farm production as spread around the market at some central or near central point. With certain agricultural commodities there is some approximation to truth with such a model, and there can be no doubt of the influence of livestock markets on surrounding farmers, of city vegetable markets on some nearby producers, or of the local market in peasant agriculture. However, some farmers find their market overseas, others avoid local markets because their product is too specialized or too large in quantity and they fear that prices will fall where there are too few buyers. Increasingly dealers are taking a larger share of the produce, and today sales are being made directly to supermarket chains or to frozen food and canning factories.

For the most part agriculture is characterized by a large number of small producing units. Individually they can have little effect on the market, and in certain situations may respond to a fall in market prices by producing not less but more. The vast number of units is managed mostly by men who have had no formal training but have learned their business by practical experience. They make a great variety of decisions in similar situations, and their decision making is greatly affected by factors outside agriculture, particularly social considerations. With so many small businesses there is little scope for the division of labour. Many farmers are unspecialized and depend on a general knowledge of managing several enterprises. In the industrial countries with more assured and affluent markets, especially for dairy produce, pigs, and poultry, the specialized small farmer is not uncommon, but even in these countries there are many small farmers unable to take advantage of new techniques or failing even to apply sufficient quantities of inputs such as fertilizer to obtain the most profitable yields. Size of farm is a crucial question almost everywhere. It decides the degree of risk-taking and may affect the proportion of specialization possible, the quantity and size of equipment and power use. In south-east Asia, for example, size of farm may decide not just whether an income may be obtained comparable with alternative incomes in industry but even whether there may be sufficient food for survival. Size is related to population pressure, to economic requirements, to land quality and to historical tradition, for certain farm sizes tend to persist in certain areas (Grigg 1963).

A final feature of agriculture is the possibility of multiplicity of products. This multiplicity has several aspects:

1. Substitution of one product for another,

(*a*) on the farm due to anticipated advantages of operation, e.g. the substitution of one spring grain for another in the system of operations, or due to a failure in the system, e.g. the failure of autumn sown grain due to adverse weather and the substitution of spring sown grain;

(*b*) in the market, where high prices for one product may encourage the purchase of a cheaper substitute, such as barley for wheat in a feed mixture.

2. Joint production in a system designed to make effective and even use of labour and equipment, to make a balanced use of soil nutrients, to reduce disease, and to reduce the cost of inputs. Traditional crop and livestock rotations of the six-course variety were intended to do this, combining spring- and autumn-sown grains, root crops for cleaning, cash and feed, seed mixtures, usually of rye-grass, rape, and oats for feed, and grazing livestock. The essence of the joint production system was maintaining soil fertility and giving full employment to the permanent labour force. The substitution of artificial fertilizers for animal manures, the development of their massive application and the substitution of machinery for hand labour have completely changed the situation. The traditional, fixed linkages of enterprises have largely disappeared in some technologically advanced situations to be replaced by rotations of convenience or even by continuous cultivation of a single crop.

3. The combination of several products for insurance against a poor market or a major failure of any one of them. The problem of risk is still important and where the producing unit is small uncertainty in production or marketing may offer the possibility of an income too low for survival or too low for the economic operation of the farm. Guaranteed price systems, government controls through official marketing boards, contracts with processing or packing factories or production for a regular market, e.g. milk, offer specialist possibilities for the small producer with reduced risk.

The farm system

The farm is operated as a firm or planned economic system, however simple or misguided that plan may be. In its use of soils, moisture, plants, livestock, insolation and slope it is also an ecological system, again planned. The study of the location of any agricultural enterprise has to take account of the existence of this two-fold system, at whatever level study is undertaken. Any regional system of agricultural production or exchange depends on the functional relationships between individual farm systems. However much we may generalize or seek detail, the actual location decision is made by individual farmers in the context of their own needs, assessments, and experience, and above all else, in relation to the system they have chosen to operate. Any enterprise distribution is therefore the result of the operation of farm systems and of the decision made by each farmer whether or not to adopt an enterprise. The significance of this may be seen in the variety of reasons which may determine the choice of a given enterprise. For example, farmers choosing to plant barley may do so because: (i) they have ideal conditions for malting barley production, (ii) they have ideal conditions for feed barley (usually light soils on which barley shows good response to heavy nitrogen applications), (iii) they wish to produce feed for their own livestock, (iv) a spring grain is needed to fit into the pattern of labour and machinery use, (v) a spring grain is needed because soils are heavy and difficult to manage for autumn tillage and planting, (vi) a spring grain is needed because autumn plantings failed, (vii) they simply prefer for no rationally expressed reason to grow barley. These different reasons, relating to different farming systems, indicate different relationships to both environmental and economic factors.

Farm systems are complex, conditioned by a multitude of internal and external factors, both fixed and variable, difficult to analyse, difficult to measure in terms of their input, operation and output, and extremely difficult to classify. All farmers make their decisions on the basis of only partial knowledge and conditioned by guesses as to the future operation of the variable factors. The farmer's problem, and the student's problem in analysis, is to separate for study or for measurement parts of the system which in normal operation are joined together and have reciprocal relationships. The preliminary model (fig. 2.1) indicates something of these relationships. As we separate the units concerned, inputs, outputs, enterprises and environment, into their constituent parts the model becomes more and more complex. For practical application thorough analysis becomes virtually impossible, partly because of the difficulties of separating joint enterprises and especially of allocating inputs between different enterprises using the same input, and partly because the immensity of the task of analysis becomes too great when compared with the value of the result. Modern linear programming techniques, for example, have to be limited in their design and application for practical farming purposes and are usually only worth applying to the study of large highly productive farms. Such techniques as network analysis, for example, evolved for the identification of bottlenecks in industrial production, are too advanced for most farmers to attempt their application (Dexter and Barber 1967, 82–3).

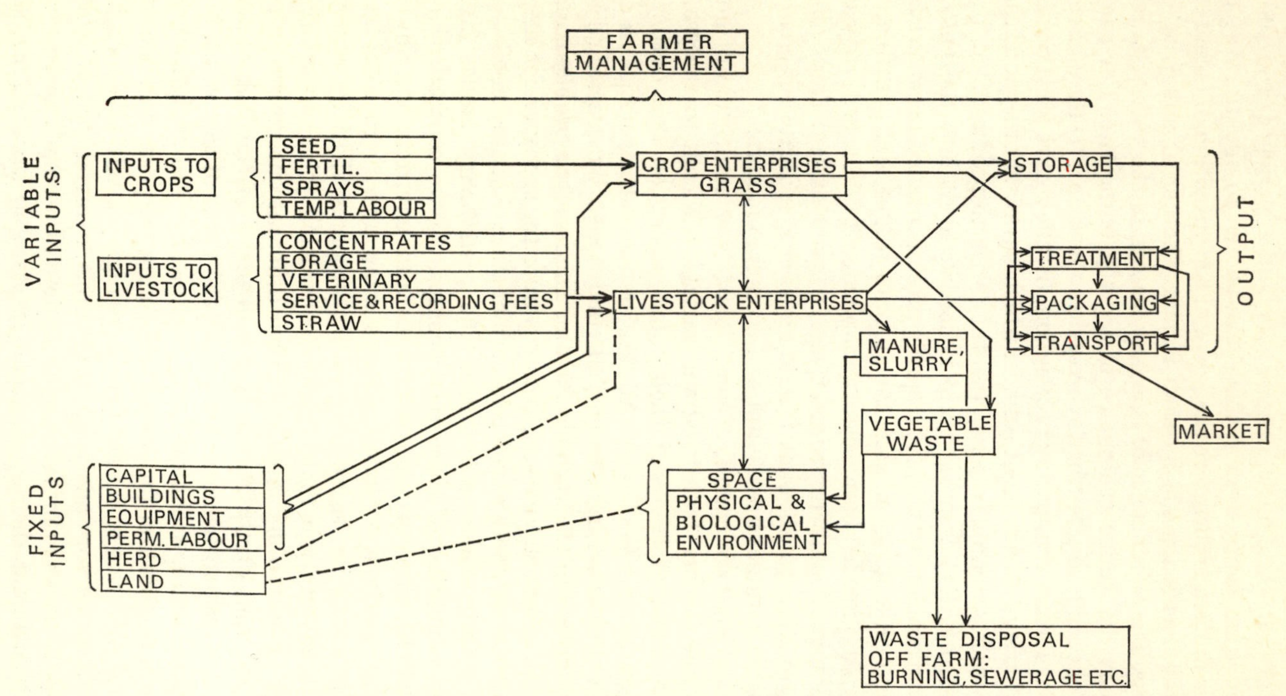

2.1 *A simple model of the basic relationships involved in a farm system.*

Enterprises

Enterprises consist of crops, grasses (sometimes classified as a crop) and livestock products. Crops consist of cereals, including the world's great starchy staples, rice, wheat, maize, barley and sorghum; roots, again including major starch producers, manioc or cassava, potatoes, yams, sweet potatoes, and also sugar beet and a wide range of root crops for animal feeding; legumes; fruit, nuts, seed crops, leaf, stem, fibre and dye plants. Some of them have industrial uses, some are for food, some are sold for industrial preparation or packing for feed, some are consumed on the farm or locally, and some have a number of alternative uses. Some crops include an enormous range of varieties, quick growing, slow growing, tolerant or intolerant of different moisture, exposure, soil and disease conditions. On some peasant farms in the tropics thirty or more different crop plants may be found, including four varieties of a single crop, all growing together in the same field. On other peasant holdings in the tropics we may also find virtual monoculture. There are few farms in Europe or North America which have as many as ten arable crops. Most have only three or four and some have only one.

Livestock include mainly cattle, sheep, goats, pigs and poultry. In some parts of the world there are highly localized domesticated animals such as the llama and alpaca, and there is an enormous number of local breeds of livestock. In Britain alone, for example, there are thirty-two breeds of sheep. In many parts of the world livestock raising is a highly specialized industry which may exist under open range conditions involving considerable wandering as in pastoral nomadism, extensively on poor grazing land as in ranching, and on specialized livestock farms either with controlled grazing or intensive 'indoor' feeding systems (zero grazing). Livestock may virtually seek their own food, find it on a managed pasture or have it brought to them. They produce meat, milk, eggs, wool or hair, hides, draught power and a considerable number of by-products. Breeds may be developed to satisfy one or more purposes, or even if specialized still make an important contribution to another purpose. Thus the main source of beef produced in Britain is the dairy herd. Draught power was formerly one of the most important livestock products in Europe and North America, and still remains important in south-east Asia. The introduction of machinery and, particularly, of the tractor has resulted in a revolution in the pattern of livestock enterprises and in the production and disposal of manures. The provision of large areas for grazing is a characteristic feature of most industrial countries despite considerable pressure on land resources for other uses. The low food productivity of livestock in return per acre when compared with that of crops is offset by the need for animal proteins and the willingness of industrial populations to pay high prices for meat and dairy produce. In considering livestock raising we must distinguish the herd, that is the breeding or productive stock, from the product or from livestock intended for sale, sometimes even described as the crop. It is important also to distinguish replacements kept to maintain the herd size. A dairy farm, for example, may contain not only cows in milk but also followers, necessary replacements for the milking herd. In most production we normally distinguish stages such as breeding, rearing and fattening, and often these are distinct specialisms operating on separate farms, livestock being sold off one farm and bought in at another as each stage is completed.

In most countries grass is still regarded as the cheapest way of feeding livestock, although in certain instances this is a false assumption and its management is often a more costly and difficult business than may at first appear (Dexter and Barber 1967, 23–4, 142–5). Grass may be self-sown or planted, permanent or rotational, grazed or mown for hay or silage. If grass is to be grown all the year round, mild winters and rain in most, preferably all, months is required or irrigation water must be supplied. The marked seasonality of temperatures and rainfall in most lands means that for part of the year grass can only be fed if it has been conserved as hay or silage. Otherwise substitutes must be sought such as feed grains, feed roots, arable by-products, or concentrates. In some cases intensive systems involving the continuous use of feed grains or concentrates are preferred. Most of the world's grasslands can only be grazed for a few months in the year and are subject to no more management than controlled grazing and burning. Improved grasslands, drained, limed, manured, dressed with artificial fertilizers and planted with the seeds of high-yielding varieties intermixed with leguminous plants, can be highly productive. There can be little doubt that most of the world's grasslands would profitably repay much increased dressings of artificial fertilizer or the application of trace elements as their present levels of productivity are very low (Davies 1960).

Official surveys or records of all these enterprises rarely distinguish their variety, the uses to which they may be put, or the particular specialized way in which they are treated. Simple plotting of wheat distribution, for example, on world or continental scales, ignores varieties, seasonality (to what time of year, if any, does our world or continental map refer?) or purpose (grain for flour, feed for livestock, grain for sale to breakfast food manufacturers, grain grown in part because of preference for an autumn-sown crop). Analysis of location factors based on the simple wheat distribution map becomes impossible or at best yields only very small profit. At the same time farm surveys are often only possible over very small areas within which generalization is difficult. Studies of the location of agricultural enterprises are therefore among the most difficult and the most unrewarding to make despite the well-established tradition for such work and its frequent citation in school and college texts.

Inputs and outputs

Inputs consist of all those items which have to be put into the system to make it work. According to the system concerned they consist of livestock, seeds, fertilizers, feeding stuffs, labour, purchases of land, machinery, vehicles, various items of equipment, buildings, fuel and power, sprays, veterinary services, and repairs and maintenance. Purchases of land, buildings and machinery may involve loans requiring interest payments. These, together with rent, are also inputs whose costs have to be deducted from sales to arrive at profit. There are also environmental inputs for which the farmer may not pay directly but which demand considerable costs involving an appropriate allocation of the inputs already listed for their control.

Inputs may be classified as fixed and variable. *Fixed inputs* are those whose effects and costs are fairly constant or change only over a period of years with changes in technology. They include items of capital equipment such as tractors and combines, together with land, buildings and permanent labour, which is the labour for which regular work must be found if it is to be used efficiently. Such inputs affect the whole system and often it is either difficult or impossible to allocate their costs to individual enterprises. In consequence it can become extremely difficult to estimate the net income of individual enterprises where two or more are contained in the system. *Variable inputs* directly affect individual enterprises and their costs can therefore usually be allocated quite easily. They include seeds, feeding stuffs, fertilizers, temporary labour and veterinary services. Such costs may be subtracted from the gross output or total income per unit area from the sales of an enterprise to arrive at the *gross margin*, a useful figure in farm management analysis as it can demonstrate weaknesses in the system, and sometimes useful for comparative purposes in the understanding of location factors.

Farm outputs can be regarded in several ways. There is first of all the general areal view of the total productions of crops, grass and livestock off the farm acreage. As some of the crops and grass may feed the livestock or even go back into the land if ploughed in, we must regard part of the production as simply used within the farm system and restrict the term output to products passing through the farm gate. Such products will normally be sold and can be measured in terms of cash earned, profitability (where measurement is possible), or, if consisting of food, of their calorific or starch equivalent value (see below, p. 106). If we regard output as not just goods sold but the total increase in value of the farm, we may have to take into account gains in livestock values and land values, and we may wish to deduct from these the costs of wear and tear, and depreciation. A farm may become a worthwhile investment even if its output of goods through the farm gate barely covers costs, if the gain in land values is greater than the gain which could have been obtained from any other form of investment. Moreover, there are 'hidden' forms of income. For example, barley grown on the farm in Britain but used to feed the farmer's own livestock is not output through the farm gate. Nevertheless it receives a subsidy payment which is part of the farmer's income.

Farm operations and management

Farm management and operations such as tillage, drainage, spraying against weeds and pests, and harvesting, are strictly speaking inputs and should form part of input analysis. From the geographer's point of view, however, it is useful to separate them because:

(*a*) farm management deals with the inputs and outputs as a whole, so that an understanding of management provides a key to the system, and is essential to an understanding of locational decision making;
(*b*) whereas when we consider inputs as such our concern is mainly with economic principles, in considering management we can use the tools of operations research and apply behavioural theory;
(*c*) farm operations involve the deployment of other inputs in a time sequence and in a spatial pattern which in part determines choice and quantity of inputs and may act as a constraint upon production.

Farm operations involve problems of season, sequence, availability or time to complete a given operation, availability of labour and equipment, and timing in relation to market forces. Most operations have distinct seasonal patterns or rhythms. Often, as in some manufacturing enterprises, the chief determinant of production is the bottleneck in the sequence of operations, or the amount of work possible in the most limited operation in the system ('limited' may refer to restriction in time available or to the incurring of higher costs or other such constraints upon production). Thus a short period available for tillage due to persistent wet conditions may restrict the tillage area in a given year and the volume of crop production, even though a far larger area could have been weeded, sprayed and harvested. Where equipment is shared or must be hired timing problems may be acute, and there is often an apparent tendency for farmers to overspend on machinery in order to make sure of preventing bottlenecks. The preferred timing of operations in relation to anticipated weather sequences will inevitably result in any region in farm commodities being offered to the market at a time of abundance when prices are likely to be low. To offer commodities at a time of shortage and high prices, however, normally involves higher costs and greater risks. In a period of generally high productivity and low market prices, more farmers will take such risks and endeavour to produce 'earlies' or will seek to minimize their costs by reducing inputs and even operating what were called in Britain at a time of agricultural depression 'dog and stick' systems.

The farm as a spatial system of operations

The farm enterprises, together with their associated inputs and activities linking them in a system of operations, have their own particular location patterns on the farm. The use of particular fields is linked to the use of others through crop rotations even if the fields themselves are fragmented as a result of ownership patterns or deliberately scattered so as to encompass different physical conditions. While numerous empirical observations have been made of the relations between the location of fields and the farmstead (Chisholm 1962), often in an historical or social context (Houston 1963, 49–79; Slicher van Bath 1963, 54–8, 164, 174), the one attempt to devise a general theory of land-use based on distance from farmstead and intensity of output has yet to be satisfactorily tested (Found 1970).

Accessibility is a major consideration. Ideally, where land is of uniform quality, buildings, or the point from which all inputs emanate and outputs have to return to be stored or consumed, should be sited as centrally as possible. Intensity of production declines with distance and this may be reflected in a change in land-use (Chisholm 1962; Morgan 1969). This is particularly noticeable when a bulky but low-value crop has to be transported, when the farmer has a considerable distance to travel to reach his fields, or when the opportunity costs on his travel time are high, as at busy times of the year, and the farmer could be much more profitably employed in other activities. For example, cows need easy access to a milking parlour and a home field or collecting yard in which to congregate before and after milking twice a day. On a sugar-cane plantation the processing plant has a fixed location. If throughput is raised to increase efficiency and sugar-cane yields cannot be increased, then the areal extent of cane to be harvested has to be enlarged. Haulage distances rise and transport costs increase, creating conflicting cost trends, so that both land and plant capital resources may be inefficiently employed. The problem with many farms, however, is not to choose the ideal layout for the system of production preferred but to choose the most profitable system bearing in mind the existing layout and the costs involved in any alteration. Cheap buildings with a relatively short life may therefore be selected.

Some farmers have to accept a degree of fragmentation or break-up of the farm unit by roads which may make the use of certain fields for dairying, for example, uneconomic or impossible no matter how suitable these fields may be in terms of their physical characteristics. The loss in this way of a portion of the farm for dairying may reduce the capacity of the remaining 'suitable' land to a size too small to support an economic number of cattle in terms of labour or machinery use and force the farmer out of dairying altogether. Similarly, land too steep for agricultural purposes other than grazing may encourage the farmer to adopt beef or sheep rearing as the dominant or exclusive enterprise as a way of making the best use of all the farm's land resources, although part of the farm could be cropped. Hill farms are often in this position, the use of the enclosed pastureland around the farms being linked to the seasonal use of upland grazings by stock which require their winter feed to be conserved from the enclosed pastures.

In some areas farm buildings are congregated for social or other advantages into villages and separated from their fields. In certain instances this creates accessibility problems which have led to a zonation of land-use around the village, normally resulting in the most intensive use being made of fields closest to the village (Blaikie 1971). The number of those members of the community who actually work distant fields also varies, these usually being the young, the more adventurous, and the most land-hungry (Moerman 1968). During sowing and harvest periods cultivators in underdeveloped economies may remain on their remote fields to reduce travelling time and to protect their crops. Mechanization can reduce the effect of distance, but usually only by making life easier rather than by reducing production costs.

3 The farm firm: problems of decision making

Two kinds of farm management decision are generally recognized. Firstly, there are planning or policy-making decisions which are concerned with such major questions as choice of enterprise and allocation of capital resources. These decisions are usually made some time in advance of their implementation and provide a 'structure plan' for the future organization of the farm. Secondly, there are organizational decisions made on a daily or weekly basis with due regard to prevailing weather and market conditions and in the advent of unforeseen problems such as livestock illness or staff absence. All land-use patterns reflect both kinds of decision. Policy making requires careful and original thought, whereas organizational decisions frequently require snap judgements. In industrial concerns different grades of executive often make these two types of decision, whereas in the farm firm the farmer is normally responsible for both. Moreover the farmer will often reach a position of responsibility on the basis of family structure rather than merit, a situation less common in industry even among small firms.

A farmer's decisions are related to his goals, of which he will usually have several, and although he may attempt to attain all of them he will normally establish an order of priorities. Any attempt to achieve in full several objectives often leads to conflict, particularly as the goals of other members of the farmer's family, or even those of his employees cannot be ignored and may even be in conflict with his own. Dalton (1967) has devised a broad three-fold classification of farmers' goals:

(*a*) physical well-being – the need to provide for present and future requirements, whether it be in terms of food production or the accumulation of wealth;
(*b*) social recognition – the achievement of status, respect, or even power within a particular community or group;
(*c*) ideological motives – these include patriotism, the idea of duty, parental and family responsibilities.

Economic geographers have been largely concerned with the farmer's desire for physical well-being and, in particular, profit maximization.

Unfortunately the term profit maximization is ill defined. In commercial economies equating total marginal costs with total marginal revenues is usually recognized as its most meaningful definition. Yet farmers and agricultural advisers do not necessarily recognize this definition and may implement farming methods that maximize returns to individual inputs. Unless agreement exists between the researcher and the farmer, the predictive power of profit maximizing models may be much reduced. In Britain, for example, the most widely employed farm-planning technique is the Gross Margin which compares enterprise margins on an acreage or land input basis. Yet many farmers consider that this technique turns attention away from the more pressing problem of returns to capital investment (Cave 1966) and plan their farms according to returns on capital investment. In fairness to the advisory services that promote the use of the Gross Margin it should be appreciated that as a simple farm-planning technique it is comprehensible to most farmers and easier to calculate than assessments on capital investments or the point at which total marginal costs are equal to total marginal revenues.

The profit maximization assumption has been criticized as unrealistic when applied to the decisions of most entrepreneurs (Simon 1959; Cyert and March 1963; Pred 1967 and 1969). It has been pointed out that a policy of short-term profit maximization often leads to a different ordering of priorities and land-use pattern than do longer-term considerations. As an illustration of this we may refer to the initial overcropping, out of ignorance and the desire for quick profits, of the American Great Plains. Farming methods aimed at soil conservation have since been implemented, and while these have reduced incomes by restricting cash cropping they should have the effect of increasing and stabilizing incomes over a longer period. Similarly, farm ownership or security of tenure encourages good husbandry, investment in farm improvements, and a concern for future possibilities, frequently absent in the farming programmes of farmers on short leases. The making of optimum decisions also presupposes complete knowledge. The complexity of the dynamic farm environment makes this in itself unattainable, and the inevitable uncertainty surrounding future weather conditions and market prices makes this assumption unreasonable. Moreover the farmer acts in response to a world of perceived circumstances which is his 'reality' leading him to seek and assess information subjectively and personally. Consequently, 'satisfaction' has been proposed as an alternative and more realistic description of goal formulation than profit maximization. This conclusion should not lead us to reject the concept of 'economic man', however, as he represents an important 'type' obviating the need for us to incorporate other less measurable goals into our decision-making models. The results obtained from optimization models can often be usefully employed as yardsticks against which the degree of sub-optimality of resource use in real life can be measured, as has been demonstrated spatially for central Sweden (Wolpert 1964).

The alternative 'satisficer' concept includes the enterpreneur's desire for satisfactory incomes, leisure time, and other social considerations at the expense of attempts to maximize profits. It also substitutes the idea of 'bounded rationality' (Simon 1959), or limited rationality, for that of 'omniscient rationality', a state in which the entrepreneur is fully informed and completely rational in his business dealings. Real-life examples of the operation of the 'satisficer' concept abound. One of the major reasons for the decline in number of dairy herds in the United Kingdom, for example, is the desire for more leisure and freedom of action than that permitted by the strict routine of twice-daily milking every day of the year. An investigation into the effects of urbanization on farming in south-east England has also demonstrated that many farmers had professional employment in London and were often 'more concerned with technical optima, such as high yields or the appearance of the farm' (Gasson 1968, 74), than with economic efficiency. These hobby-farmers often preferred to view the farm as a home or a place of recreation rather than as a source of income. One model of the farm firm embodying non-profit maximizing assumptions has been successfully developed and tested for family farms in Indiana (Patrick and Eisgruber 1969). In this, the desire for leisure, the needs of the family – especially those of the children – risk avoidance and the wish to own land, as well as raising farm income, are explicitly recognized. The success of this study suggests that in a highly developed and competitive farm economy with well-educated farmers and a rapid diffusion of information, the effects of a limited number of non-economic motives can be ascertained, measured, and evaluated, and that these are of significance to the decision-making behaviour of a majority of farmers.

Farmer expectations

Farmers' expectations of future events are largely based on their experience of the recent past. Events from the distant past are only vaguely recalled unless of prolonged or of catastrophic proportions. Older farmers, in parts of the United States, remember rampant soil erosion on the Great Plains (Saarinen 1966) and some British farmers remember the 1930s agricultural depression. Exaggerated recollections of events like these may lead to excessive caution in contemporary decision making. More important, however, is the fact that most farmers project recent crop yields and prices into the immediate future with only occasional minor modifications. Longer-term anticipation is nearly always tempered with pessimism and caution and, unless drastic price changes are envisaged, rarely receives the considered attention it warrants. The limited contact between farmers and retailers common to many developed agricultural economies means that farmers' expectations are usually based on supply rather than demand information. The farmer feels more able to assess the state of the harvest than the changing whims of the housewife and frequently assumes that demand is stable, which is true for many staple foods over the short term, and that price fluctuations result simply from variations in supply. However, with the development of vertical integration and a more quality conscious consumer this outlook is bound to change (see chapter 6).

The farm environment

Farms function within more general social and economic systems. Falling profits resulting from government legislation over product prices may force small farmers out of business, lead to greater specialization, or to a reduction in inputs on larger, more viable units. Similarly,

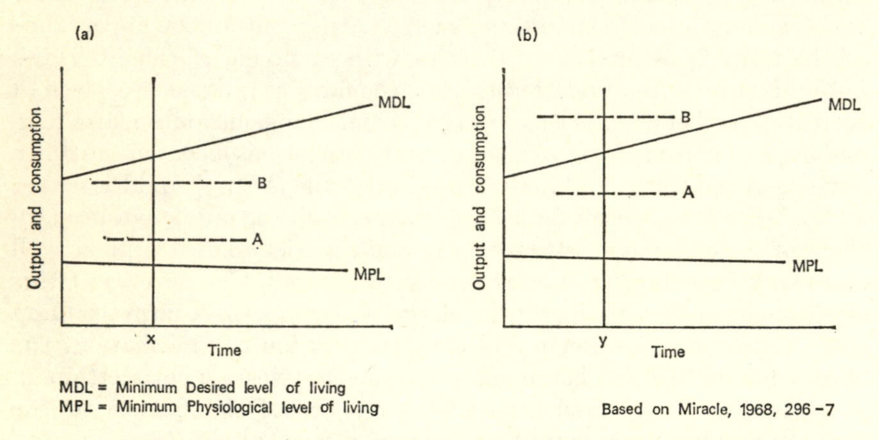

3.1 *Relationship between goals and output levels and its effect on willingness to innovate.*

In (a) at time x farmer A is unwilling to innovate in case this leads to a fall in output below the MPL while farmer B is prepared to innovate as this may lead to an increase in output above the MDL.

In (b) at time y farmer A is in the same position as farmer B in figure (a) while farmer B is now unwilling to innovate in case this leads to a fall in output below the MDL.

urban development may result in a complete reappraisal of potential market outlets, but only if the farmer has the resources to adapt his system of production to the new opportunities offered. In many rural communities change is avoided or even resisted and if conservative attitudes prevail an individual may be unprepared to innovate, for innovation involves risking not only capital and profits but also status and respect. Indeed, the whole farm environment tends to militate against change as it cannot always be assumed that it is the successful farmer who will be prepared to search for and invest in a new system of production, irrespective of the apparent financial gain it would bring (see fig. 3.1). This is because any new venture incorporates increased risk in an already uncertain environment, and as a small businessman with limited resources the farmer cannot afford to fail. Low profits over a number of years reduce the opportunities for change as the farmer may not have sufficient resources left to do so, particularly as many medium- and long-term farm investments are not easily liquidated without considerable loss. This situation is exacerbated where the production process itself takes a number of years to realize any profits, as with tree crops for example. In conclusion, therefore, the decision-making situation of the farmer is often one of inability to act, and this predicament has been summarized by Gasson, who, having analysed the occupational immobility of small farmers, commented that 'as a broad generalization, it appears that small farmers who COULD move out of farming WILL not, whilst those who WOULD move CANNOT' (Gasson 1969, 284). An additional point here, of some importance, is that small farmers tend to be more capital-intensive in their operations than large farmers in order to provide minimal acceptable incomes. In consequence they frequently lack flexibility in the face of technical change.

Time

Time change is implicit in the decision-making process, but it is rarely considered explicitly (exceptions to this include Thornton 1962; Harle 1968). This is indeed surprising as the rate of change in the social, economic, and technological environments surrounding the farmer today is more rapid than it has ever been. As a result short-term organizational decisions are gaining in significance and attempts to isolate a discrete time-sequence for a particular decision are becoming increasingly difficult. Most farming decisions have to be made by a particular date, often dictated by the seasons, and the amount of time available for the farmer to make a decision will affect the thoroughness with which he can collect and assimilate information. As he rarely has all the required information to hand he has to instigate a search procedure involving, for example, contact with his advisory officer or discussion with a friend or even listening to farming programmes on the radio or television. All this takes time and often means that decisions have to be made with inadequate knowledge as postponement means a twelve-month wait where an enterprise's economic and husbandry dating tolerance limits are set by an annual, biological cycle (Duckham 1963). More generally, the dynamic nature of the whole environment requires a rapid readjustment on the part of the farmer to changing circumstances and as a result our whole emphasis in the study of decision making should perhaps be re-examined as it is more important for the farmer 'to know how to respond to changes in conditions than how to behave optimally in any single set of circumstances' (Dorfman 1968, 56).

The availability and use of information

It is important to distinguish between the 'awareness' of information on the part of the individual and its 'use'. While these two states may be synonymous, particularly as the farmer will not invoke a search procedure unless he plans to use the information obtained, any farmer will inevitably acquire 'redundant' information from his day-to-day activities. This distinction is significant as the description of innovation diffusion, for example, is dependent upon 'use' rather than 'awareness' of information. Both the amount and quality of information, quality being defined as the level of confidence placed in it by the receiver, affects its acceptance and use. The many sources of information available to the farmer may be classified into two groups:

(*a*) sources external to the agricultural society, e.g. agricultural advisory services, research centres and mass media sources;
(*b*) sources within the agricultural society based on inter-farmer personal contact.

Information obtained from sources external to the agricultural society is usually treated with greater circumspection by farmers than information received from other farmers, although these sources often provide the 'awareness' if not the acceptance or use of information. As a way of obtaining new information inter-farmer personal contact is of least significance to innovators as they obtain most of their information from outside the farming community. On the other hand, it has been demonstrated that personal contact between farmers is a useful basic assumption in the simulation of innovation diffusion (Hägerstrand 1953; Bowden 1965), indicating its significance for many farmers. The innovator, however, should not be viewed simply as an early adopter of a new technique as he often embodies a leadership function within the farming community through which he not only informs other farmers of new ideas but also influences their acceptance of them.

Acceptance of information depends not only on its source but also on its content, particularly its 'newness'. If the idea merely extends an existing technique or the farmer has experience of an allied system of production he is less likely to reject it than if the concept is totally new. Similarly, he is likely to require less information about it before he accepts it. At the same time the farmer's personal characteristics, education and outlook influence the way he searches for and acts upon information received. It is well known that young farmers, and in particular those with higher education, are more prepared to take risks in order to try out new ideas than elderly farmers who have lost the mental and physical drive necessary for experimentation. 'Urbanization', defined by Emery and Oeser (1958) as the degree of contact with urban life experienced by the farmer, is also recognized as a potent force for a more liberal attitude towards new ideas, and this implies a significant spatial influence on the diffusion of information, although at present, little is known of the information flow resulting either from the daily movement of part-time farmers in and out of urban areas or from the migration of full-time farmers from more to less urbanized regions.

At least one ambitious attempt has been made to systematize the role of information in decision making, through the construction of a 'behavioural matrix' (Pred 1967 and 1969). However, the two axes of the matrix – availability of information and ability to use information – are interdependent as the information sought by the decision maker and hence 'available' depends on his perception of the environment, which in turn is related to his ability. Although the concepts behind the axes are valid in themselves, their interdependence means that it would be impossible to locate with accuracy the cell in the matrix to which each decision maker belongs.

Uncertainty

Although at the industry level the mean annual variation in agricultural income is no greater than that of other sectors of the economy (Jones 1969), at the farm level variation can be considerable. The analysis by Jones reveals that the variability of farm incomes in Britain is twice that for firms in other sectors of the economy, despite the Government's policy of guaranteeing minimum product prices. In south-west England, although the average annual farm income is £26·70 per hectare the 95% probability range for income in any one year extends from £4·50 to £48·00 per hectare and the annual average deviation from the mean is ±50% (Langley 1966).

Variable future conditions, whether they be income levels, yields or rainfall amounts, can be separated into two types. Firstly, there are those future conditions whose previous occurrences have been recorded and the probability of their future occurrence can be predicted. The greater the number of recorded occurrences the more reliable will be the probability value calculated. In theory, decisions taken by farmers in the developed world, where records of market prices and weather phenomena have long been kept, should fall into this category. In practice, this is frequently not the case as most farmers have neither the ability nor the inclination to obtain detailed records and derive the appropriate probability values. They prefer to calculate on rule-of-thumb bases using data from the recent past. Moreover, as the environment within which decisions have to be made is dynamic, the relationships between its aspects are continually changing. For example, the relationship between rainfall and crop yield changes with technological advance and its significance to farm income with changes in market price. Secondly, farmers may be faced with completely novel situations, at least to them, when the probability values for particular outcomes cannot be calculated. These occasions can take the form of revolutionary technological developments, such as the introduction of the tractor, or can result from decisions beyond their control such as that of a local authority to build a new housing estate either on or in the vicinity of their farms. Yet absolute uncertainty rarely exists as few situations have no precedent or referent and the farmer can nearly always acquire relevant information if he searches extensively enough for it. Few farmers gamble on totally new methods of production and it is more usual for farmers in uncertain situations to gain a sense of security by copying the actions of their neighbours. However, the suggestion that all decisions should be regarded as unique events, which means that, by definition, their possible outcomes cannot be related to any probability distribution (Renborg 1962) would seem to be an unduly pessimistic assumption. On the other hand, the response of individual farmers to uncertain situations is unpredictable, although they may react in aggregate in a predictable manner. A major difficulty surrounds the fact that even where statistical information permits the calculation of 'objective probabilities', the farmers' interpretations of these may be very subjective, depending upon their personal perception of the uncertainty concerned. As a result, subjective and objective probabilities may differ drastically and the predictive values of models based on objectively determined probabilities be called into question.

This is a criticism that can be made of Game Theory which having derived the probability of future outcomes objectively from statistical records, then relates them to the decisions of individual entrepreneurs. Agricultural geographers have used the method primarily to estimate the effect on decision making of uncertain future weather conditions but economists have also employed it to measure response to uncertain and variable future market prices. Although numerically sophisticated, the principles of Game Theory can be simply outlined. The method assumes that a matrix, known as a pay-off matrix, can be constructed to show, for example, expected returns from alternative cropping systems under varying weather or market conditions. The pay-off matrix described by Harvey (1966) provides a suitable example.

This pay-off matrix assumes that four distinct weather types occur with equal frequency and result in the incomes indicated for the three alternative cropping systems (systems cannot be mixed). Choice of cropping system can then be made according to a number of decision models (Agrawal and Heady 1968). The best known of these is the 'maximin' model in which the farmer selects the cropping system that realizes the highest minimum income under any weather condition (cropping system A), irrespective of average income over all-weather types. Such a procedure would be followed by a farmer who needs a

	Weather Conditions				
	1	*2*	*3*	*4*	*Average*
Cropping System		*(income in £)*			*(£/year)*
A	450	550	500	500	525
B	700	300	900	340	560
C	0	1000	0	3000	1000

given food output or income per year, in this case £400, in order to survive, and is a particularly appropriate model for small, subsistence farmers. Those farmers who can withstand years of disastrous failure would choose system C since this maximizes returns over all-weather types in the long term. As individual farm circumstances and individual perceptions of an uncertain future are bound to vary, the use of this normative method, despite the apparent success of Gould (1963), either to describe land-use patterns or predict their future changes can be questioned. Until more information is known about the subjective assessment of probabilities by farmers individually or in aggregate, or the very generalized model used by Gould is tested for more than one West African village, the possibility that the close fit in the latter study between reality and the predicted land-use pattern occurred by chance should not be ignored.

4 Agricultural activity and the physical environment

Any attempt to describe individually all the relevant aspects of the physical environment would in all probability underestimate their interdependence and be incomplete. Furthermore, the influence on agriculture of individual aspects of the physical environment, or even complete environmental systems, cannot be meaningfully evaluated outside the context of a given production system. For example, any discussion of frost risk would have little value unless it was related to a particular enterprise, at a specified date, and in a known location. Consequently, the physical environment is best outlined in terms of its more important ecological characteristics.

Firstly, spatial variations in the physical environment place limits on the distribution of plants and animals, although the actual distribution will depend on man's willingness and ability to ameliorate physical limits. Klages (1949) has described the relevant physical limitations of the world's more important domesticated plants and outlined the concept of the 'optimum' environment, or that environment in which the plant's maximum yield can be most easily obtained. Secondly, the interdependence of all aspects of the environment affects their individual significance. The effects of exposure on animals, for instance, are exacerbated by rain, and the availability of moisture to plants depends not only on its amount and timing but also on the texture, structure, and organic matter content of the soil. Thirdly, the susceptibility of a plant or animal to a particular environmental stress depends on its stage of growth. Comparatively little is known about this, but, generally, environmental stress is considered to be more significant in the early stages of a plant's life cycle (Milthorpe 1965). In some cases the relative significance of different environmental characteristics changes with increasing plant maturity. For example, the final yield by weight of a sugar-beet crop is more dependent on rainfall during the early development of the root, while the final yield by sugar content is more related to hours of sunshine in the later stages of the crop's maturation. Fourthly, the environment's variability in both time and space, at whatever scale measurements are taken, is one of its important characteristics. Seasonality and spatial variations in land quality in particular need to be taken into account.

The biological nature of the production process provides the link between the physical environment and the spatial organization of agricultural activity. Many agricultural geographers have tended to assume that its influence is readily observable and measurable. However, their research has often been based on naïve assumptions which anticipate deterministic, rather than probabilistic, relations between agricultural production and the physical environment, as well as underestimating the complexity of the interrelations between individual aspects of the agriculture-environment system. Moreover, the entrepreneur's perception of the limitations imposed by the physical environment will be determined by his social and economic environments.

Agricultural systems and the physical environment

THE ECOSYSTEM APPROACH

The ecosystem concept has been successfully employed as a method of approach to the analysis of land-use patterns, and agricultural systems in particular (Simmons 1966). It has been pointed out by Harris (1969) and Geertz (1963) that primitive agricultural systems have similar structures to natural ecosystems, differing only in that self-sown plants and wild animals are replaced by domesticated varieties and breeds. As managed systems, however, the land-use patterns of agricultural systems depend not only on the physical environment and plant/animal relations but also on social and economic considerations. Generally, natural ecosystems contain a larger number of plant and animal species than the more specialized agricultural systems that take their place. Consequently, natural systems are more stable and their biological productivity may in certain circumstances be higher. Soil resources are more easily exhausted by the more specialized and concentrated demands of agricultural activities, especially arable farming, than by natural vegetation cycles, and the degradation of soil resources can lead to soil erosion and the progressive breakdown of the whole system. Studies in the underdeveloped world reveal a high degree of ecological perception on the part of the indigenous cultivator, as indicated by the farmer's ability to select those systems of production and techniques of management that make the most efficient use of available environmental resources (see, for example, Moss and Morgan 1970). Only when circumstances beyond the control of the individual cultivator occur, such as rapid population increase, have traditional farming systems led to an over-exploitation of resources. In more advanced agricultural societies farmers, in their search for greater profits, have often shown a lamentable lack of ecological perception. Established systems of farming have often been introduced wholesale into physical environments which have proved to be totally unsuitable. The introduction of continuous grain production into parts of the Mid-west of the United States is an often quoted example. Here farming practice had to be reorientated on conservation principles in order to put the system back into ecological 'balance', and although this led to some reduction in immediate financial gain, it represented the most likely means of maximizing long-term profits. Here again, however, we should note the increasing adoption of scientific soil management principles and the development of a considerable fertilizer technology.

Except where soil erosion has been a major problem, the ecological approach to the understanding of the functioning of agricultural systems has had few adherents. Nevertheless, in societies where food production is the most important goal an ecological approach can encourage the development of methods of estimating the greatest biological output attainable from alternative crop mixes on particular sites without an over-exploitation of the environment's resources. However, there are difficulties. To assess productivity accurately flows of energy and movements of matter within the system have to be measured, and this is not easy even within simple agricultural systems (Macfadyen 1964). Moreover, cultural preferences may not coincide with biological optima. In Malaya, for example, the introduction of a new variety of rice by the Japanese was resisted as it did not suit the palate of the Malays, although unlike indigenous varieties it could be double-cropped because of its shorter growing-season (Rutherford 1966).

If the major aim of the farmer is to earn an income rather than feed his family directly from his own produce, enterprises realizing at least the minimum desired income have to be incorporated into the farming system. Unfortunately, responses to economic inputs are delayed by the biological nature of the production process, resulting in inefficiency in resource use and uncertainty in decision making. In most situations the physical environment reduces enterprise choice either by prohibiting the growth of certain crops altogether or by reducing their levels of output to an unprofitable degree, although its actual economic influence will depend on the limitations it imposes on the economically preferred enterprises, available technology, and the ability of the farmer. Spatial variations in economic phenomena and the economic significance of the physical environment, rather than ecological considerations *per se*, will therefore dictate spatial variations in agricultural activity. In this context analyses at the farm level have a particular value as farmers do not evaluate environmental resource in isolation from particular production units and their decisions concerning environmental limitations must be judged accordingly.

TECHNOLOGICAL CONTROL OVER THE ENVIRONMENT

In commercial economies it is not only the availability but also the cost of implementing new technological developments that is important. The cost must relate to the value of the enterprise as well as to its sensitivity to the physical limitations in question. As satisfactory economic returns have to be obtained from costly technological innovations, the economic importance of these limitations will depend on the overall profitability of the farming industry. In Britain, for example, their importance should not be underestimated. In the last decade the real price for many farm products has remained static while costs have risen and profits per unit of output have fallen. Only greater output through a more intensive use of land has maintained farm incomes, and consequently the farmer still requires, despite technological advances, reasonable weather and few soil limitations to implement his more ambitious cropping programme successfully.

Undoubtedly, the most important benefit to agriculture of technological advance has been to increase the range of enterprise choice within a given physical environment, and this has been achieved not only by ameliorating its effects (Sewell 1966) but also by improving husbandry techniques. Attempts are continually being made to reduce the uncertainty associated with weather variability and to extend the growing season by such means as cloud seeding and irrigation. Fertilizers, drainage systems and mulches have reduced soil limitations and more powerful machines, new buildings, improved plant varieties and animal breeds have increased output. Total control over the physical environment is an ultimate possibility, and in small measure this has already been achieved with, for example, soil-less cultures for horticultural purposes and controlled environment houses for broilers and battery hens. Farming completely under glass or plastic sheet is no longer impossible when its use for horticulture has already extended to units under cover as large as 20 hectares, using glasshouses of 1·3 hectares each (Cooper 1970). Other farmers zero-graze dairy cattle and intensively rear barley-beef, the cattle never leaving the buildings. The major management problem today is not so much the technical difficulties associated with these production systems as the most effective use of the long-term capital investments tied up in them.

This degree of environmental control has completely removed a major locational determinant for some systems of production. Large-scale livestock producers, for example, now find that transport costs have increased, at least in relative terms, within their cost structures. Some 'farms' have therefore been relocated near the market, such as the town dairy in California and Japan, where the difficulties of transporting liquid milk exceed those for feed, now consisting mainly of grains and concentrated proteins (Gregor 1963*b*). Other production units have been located at the source of feed, as is the case with intensive beef fattening in the Colorado valley, where sugar-beet pulp, a bulky by-product of locally produced sugar beet, is fed to cattle in buildings. The removal of environmental limitations has led to other locational trends. For instance, some horticultural crops previously grown under glass in the Lea Valley can now be produced on a field scale in the open in such favourable locations as the Fens (Gasson 1966*a*).

PERCEPTION OF THE ENVIRONMENT

The farmer's use of his resources will depend on his perception of them rather than any objective measure of their characteristics. His perception of their value for alternative production systems will depend on his background, information and ability; and his yardstick will be based on his past experience of them. For example, does a farmer on light land view a medium silt soil as heavy land? Similarly, does a farmer on clay land view medium silt soil as light land? More interestingly, do both farmers view the productive capacities of a medium silt soil as the same? It would be tempting to conclude 'no', but conclusive evidence would be hard to obtain since any farmer's evaluation of land quality would also be determined by his perception of the farm's other resources and the returns to be obtained from them as a whole. Furthermore, it is well known that able farmers obtain greater returns from the same land type than their less able neighbours, often through a more intensive use of the same land resource.

Environmental perception studies have been reviewed by Brookfield (1969) and represent a natural development of earlier cultural and historical geographical research. Studies by agricultural geographers have focused on the reactions of farmers to risk aversion, particularly flood risk (Burton 1962), and drought hazard on the Great Plains (Saarinen 1966). Saarinen found that although experienced farmers were aware of and could accurately assess drought risk, they tended to underestimate the frequency of drought years and overestimate the number of good years. He was able to conclude that farmers in the most arid areas with the most drought experience and whose farming operations were most vulnerable to variations in the weather most accurately perceived drought risk. This complements the conclusions of Tefertiller and Hildreth (1962) who noted that farmers operating in very uncertain Great Plains' environments attempted to predict the type of forthcoming season and adjusted their management practices accordingly. If low rainfall and as a result low yields were anticipated, fertilizer inputs were reduced and only minimum cultivations carried out, on the grounds that if the crops failed only limited inputs would then have been wasted.

SPECIFIC MANAGEMENT PROBLEMS

The physical environment presents two management problems to farmers worthy of further consideration, weather uncertainty and seasonality.

Weather uncertainty

Enterprises vary in their susceptibility to adverse weather conditions, and therefore the uncertainty involved in their production and sale. Most susceptible are those products that have to reach the market at exactly the right time to command the best price, for example, early potatoes and strawberries, or have to be harvested on a particular day to avoid deterioration in quality, such as peas. Most enterprises have critical husbandry dating schedules, or those dates by which particular husbandry operations have to be completed if yield loss or, in extreme cases, total crop failure, are to be avoided (Duckham 1963; 1967). If these get out of step early in the husbandry programme a whole season's cropping can be affected. Arable crops are weather sensitive as they entail numerous field operations, while livestock production, unless within a controlled environment, can be directly affected by a response on the part of the animals themselves to extreme weather conditions and indirectly by reduced supplies of forage.

Farmers can combat uncertain weather conditions by purchasing controlled environments, planning flexible cropping programmes and diversifying production systems. Not only can enterprises affected by local weather hazards be mixed with those that are less susceptible, but crops requiring different weather conditions can be grown. This leads to conflicting weather requirements, for although late summer rain hampers grain harvesting in Britain, it benefits grass growth. Uncertainty can also be greatly reduced by simply increasing yields in poor years. This raises the minimum expected income, so important for the survival of many farm firms. In the United States, for instance, new grain varieties have not increased crop yields in good years but have substantially improved them in poor years (Shaw 1964).

Seasonality

The demand for food by man and animals alike is largely aseasonal, while the economic and biological inputs and outputs of farming have a marked seasonal pattern. The seasonal output pattern places a premium on efficient storage facilities, yet these alone cannot remove the annual experience of hunger in many underdeveloped countries. In commercial economies it is the flow of cash in and out of the industry that leads to management difficulties. The most serious problems arise when the need for substantial labour input coincides with a period of hunger, or the need for credit with a period of debt.

Geographers have noted that the co-ordination of grass output with livestock demand is a production problem faced by many temperate mid-latitude livestock farmers (Curry 1962). The demand for food by intensively grazed dairy cows is both substantial and largely aseasonal whereas the rate of grass growth varies considerably between seasons. The farmer can employ a number of management techniques that will solve this problem, the most obvious of which are the conservation of surplus spring grass for winter feeding, the sowing of several grass varieties with different seasonal growth patterns and irrigation during dry periods.

Seasonal variations in levels of production lead to varying market prices, particularly for perishable but staple foods for which there is an aseasonal demand. For example, in Britain it is cheaper to produce milk in summer than in winter. To deter seasonal variations in supply the Government pays a higher price for winter milk with the result that only 52% of total output is now produced between April and September. Labour inputs vary considerably round the year for most enterprises

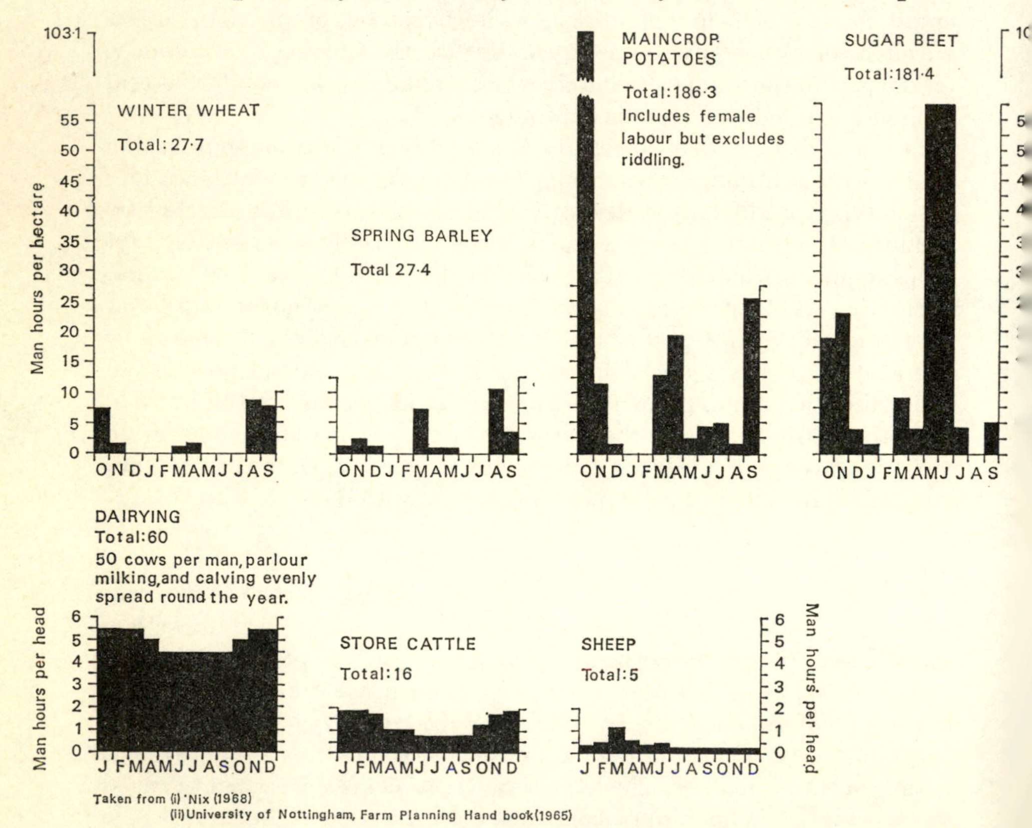

4.1 *Labour inputs per month by enterprise, Britain mid-1960s.*

(fig. 4.1) with the result that many farmers employ a mixed system of production in order to keep their labour fully employed. Yet seasonal unemployment remains on most farms although this can be reduced by employing casual labour during peak work periods, and in developed countries by substituting machines for labour. However, this in turn often leads to the inefficient use of capital as many machines are either only in use for three or four weeks a year, such as combine harvesters, or in very varying amounts throughout the year (for a detailed discussion of seasonality, see Duckham 1963).

5 Land, labour and capital

Economists have found great difficulty in defining the factors of production and, for some, the matter is still unresolved. Consequently, the following definitions are intended to provide merely a guide. 'Land' is viewed as area with different natural attributes. It realizes different rents and varies in purchase price. 'Labour' represents all human services other than decision making and 'capital' the non-labour resources employed by the farmer. It is the varying form that capital takes, such as buildings, machinery, livestock or cash, that is of particular concern in its assessment. Management, often termed 'enterprise' organizes the use of land, labour and capital and is itself generally recognized as a factor of production, but as this has already been discussed in the context of the farm firm it need not be re-examined here.

Land and labour are common to all agricultural systems. In a completely subsistence economy where all agricultural output is consumed by the producers and money does not enter the social system even through the sale of non-agricultural products the role of capital may be discounted. In some peasant economies the farmer receives small amounts of money and some of this may be reinvested in the farm business. The concept of capital accumulation has therefore some meaning for the peasant who is in the process of becoming an entrepreneur, for example, the Ghanaian cocoa farmer, and may influence his decisions. On plantations considerable capital investment is tied up in processing plant, and the importance of this as well as labour as an input (labour represents 60% of total variable costs, Courtenay 1965) make high returns to these inputs a major goal in efficient plantation management. In advanced commercial economies farming as a way of life is retreating before farming as a business. The development of agribusiness with its capital intensive and vertically integrated production systems (see chapter 6) is reducing the relative significance of land and labour and emphasizing the importance of capital deployment and marketing skills. Even in some peasant systems a managerial revolution is being encouraged by government advisory and extension services and new classes of 'master farmers' or commercial growers trained in modern methods that can be applied to larger holdings are being created (Kay 1969, 511–14; Takes 1964; Yudelman 1964, 140–3).

The presence or absence, amount, quality and price of each factor of production varies spatially, affecting the relationships between them and their deployment on individual farms. These spatial patterns are not static, labour and capital being geographically mobile. The use of each production factor will not depend solely upon its availability. It will be influenced by whether technological, economic and social circumstances permit the substitution of one for another, and this in turn will be affected by their degree of divisibility. The farmer may find, for example, that the smallest piece of equipment that he can buy has a capacity of twice that required. He cannot purchase half a machine, and while the machine may be essential for the production of a new and lucrative crop or for a reduction in the degree of risk involved in the growing of an existing one, its purchase may make one of his employees partially redundant.

Labour

Theoretically, labour is a highly divisible factor of production as its input may be varied in several different ways:

'1. In units of one whole-time worker.
2. By the use of part-time personnel.
3. The regular labour force may work overtime or short-time.
4. Casual labour may be employed.
5. A firm that has excess work in hand may sub-contract some jobs . . .' (Chisholm 1966, 113).

Equally, contractors may be called in to carry out tasks beyond the capacity of the existing work force. In practice, however, labour input is less divisible than this classification would suggest, particularly on small farms with few workers. In such firms the manager has little room for manoeuvre as each employee represents a sizeable proportion of the total work force and although the availability of part-time and casual workers is usually considered to be greater in agriculture than other industries, in developed economies these two categories of worker are declining rapidly in numbers. Furthermore, although the working of overtime in agriculture is usual, the extreme variability in the length of the working week is an important contributory factor to the drift of labour from the land in Britain (Cowie and Giles 1957), and farmers may have to reduce overtime working in the future if they are to keep their existing labour forces.

The level of labour input per unit of land or capital depends on its availability, its cost and the need to achieve a given level of output. Systems of production vary in their total labour requirements as well as in the seasonality of their demands (see chapter 4). Generally speaking, the marginal productivity of labour declines with increasing inputs (fig. 5.1) and it is quite normal for the farmer to obtain a lower marginal return for some of his effort than he would obtain from being employed on another farm. This phenomenon not only occurs on small, subsistence holdings in underdeveloped economies but is also a feature of many hill farms in Britain for example. It is possible for the marginal return to labour to fall to zero although this rarely occurs even in overpopulated

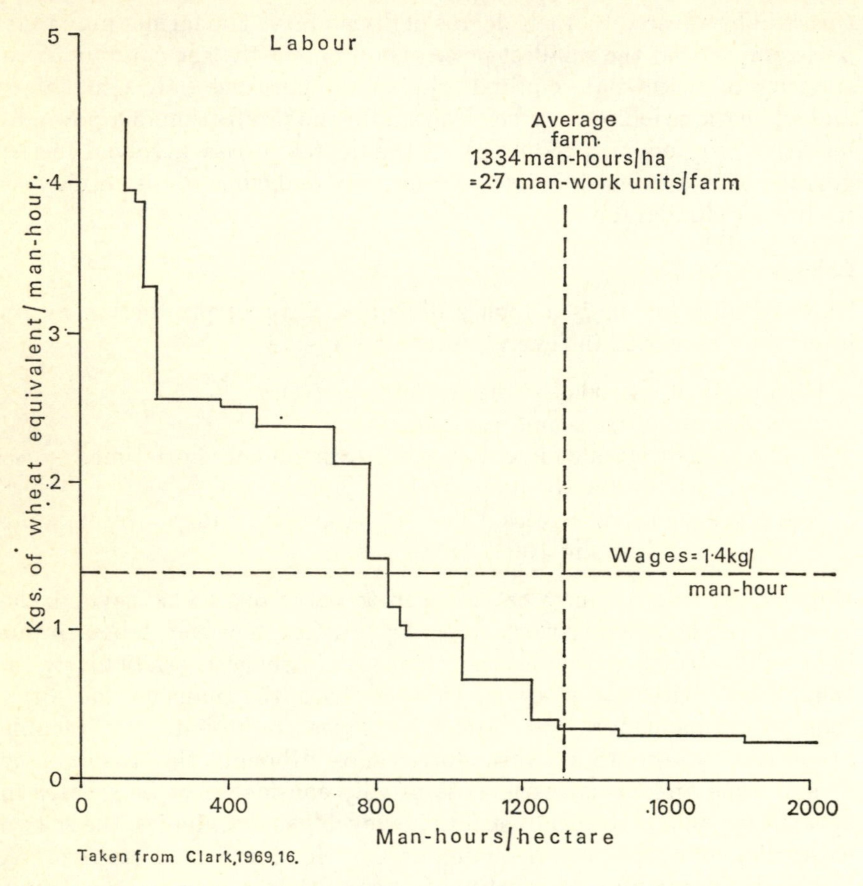

5.1 *Marginal productivity of labour: an example from northern Greece.*

areas. More prevalent, and almost as serious in subsistence economies, is that part of the marginal return to the farmer's effort will be less in calorific value than the energy expended in obtaining it.

Labour is both occupationally and geographically mobile and both forms of mobility are frequently associated. Generally, farmers are less occupationally mobile than other workers. British surveys reveal that over 70% of farmers' sons intend to stay in farming, although the proportion declines as the size of farm business falls (Gasson 1968).

The low occupational mobility of farmers is emphasized by the tendency for farmers' sons to avoid any consideration of white-collar employment and to accept parental rather than school careers advice. Inevitably, the desire to avoid occupational mobility decreases geographical mobility, and the geographical mobility of labour leaving agricultural employment is particularly low in developed economies. This presents regional planning problems wherever the number of agricultural workers requiring occupational redeployment is substantial. Moreover, the number of migratory agricultural workers, such as fruit and potato pickers and sugar-beet hoeing gangs, is also declining rapidly in developed economies. Mechanization has reduced the need for them and alternative industrial employment opportunities, providing regular and secure incomes for sedentary workers, are now more widely available in rural areas than hitherto.

In the United States and Britain there is current concern over the rapid loss of farm labour. In Britain in 1950 there were 843,000 full- and part-time agricultural employees, but by 1966 this number had fallen to 488,000. The National Plan (1965) estimated a further decline of 125,000 between 1965 and 1970 and Cowling and Metcalfe (1968) have revised this figure upwards to 140,000. Young farm-workers are the least satisfied with their working conditions and make up the majority of those leaving the land. There are two basic reasons for the decline in numbers. Firstly, industrialized nations offer alternative and financially attractive employment. In Britain, the average weekly earnings of an agricultural worker are £14·97 for a week of 50·1 hours, while the national average for all employees in all industries is £20·25 for a 46·6 hour week (Black 1968). This disparity is put further into perspective by the large proportion of agricultural employees earning less than £15·00 per week (approximately 60%, see fig. 5.2), whereas only 16 out of 130 industries in Britain have more than 30% of their employees earning less than £15·00 per week. All surveys of agricultural labour in Britain state that the loss of farm labour is primarily due to the wage differential. This factor becomes most significant where alternative employment opportunities exist, the loss of farm labour being greatest on the urban fringe where agricultural wages are also highest. Many explanations for these low wages have been suggested. The supposed fringe benefits of the farm-worker are often quoted, in particular, a free cottage, free milk, eggs, potatoes and firewood. Not all farms provide these and whether a tied cottage, i.e. a home tied to employment with a particular farmer, can be considered a benefit or a liability depends on whether you are the farmer or the farm-worker and whether there is an available supply of alternative cheap housing. In any case, the tied cottage must contribute to the geographical and occupational immobility of British farm-workers. Trade union activity is also traditionally less militant than in other industries, probably as a result of the difficulty of organizing co-ordinated action due to the members' scattered work-places.

Secondly, whereas most agricultural systems provide a variety of work, although under working conditions that would not be tolerated by many industrial workers, increased mechanization means that farm work is becoming more repetitive and boring. Even more significant, certain tasks have to be carried out every day of the year and attempts

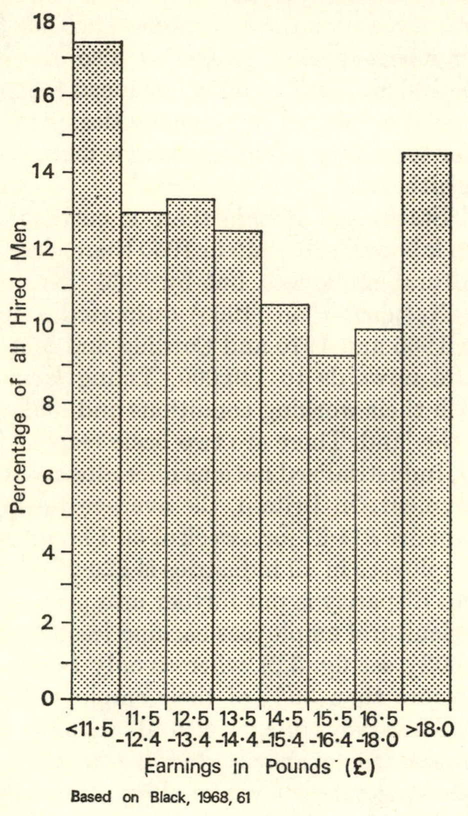

5.2 *Range of earnings of British farm-workers (1967).*

to introduce shift-milking and a five-day, forty-hour week on dairy farms in California (Gregor 1963*b*) are very much the exception rather than the rule. The greater leisure opportunities of the industrial worker will become all the more apparent as incomes rise and leisure assumes increased significance.

The underemployment of labour in agriculture is most frequently associated with underdeveloped economies where few job opportunities exist outside agriculture and is particularly marked in areas with pronounced seasonal rhythms in productivity. It should not be overlooked, however, that some advanced commercial economies have a similar if more limited problem created by the indivisibility of inputs and the resultant uneven substitution of capital for labour at the individual farm level. In Illinois, for example, 45% of those leaving agriculture claim to have been underemployed while on the farm (Guither 1963) and it would not be unreasonable to assume that other agricultural workers in the United States are likewise underemployed.

Despite much evidence of disguised, as well as evident, underemployment in agriculture in many underdeveloped countries, sweeping generalizations as to the seriousness of the problem should not be made lightly. In India, for example, regional levels of underemployment are very variable and the proportion of small farms cannot always be equated with underemployment. Small farms are often concentrated on the best land where more labour intensive systems of production are possible, and in at least one case, Kenya, labour rather than land has been shown to be the primary limiting factor to the profitability of peasant cash-crop farms (Clayton 1964). Both regional and local variations in labour underemployment are dramatic and this spatial component deserves greater attention than it has so far received. More information is available on the extreme seasonality of labour input in some farming systems. This may camouflage the real picture, emphasizing underemployment during much of the year and minimizing the small margin between labour required and labour available during the short but critical sowing and harvesting periods.

Capital

Most farms are not public companies, although companies operating plantations or the processing and marketing of agricultural produce, may be quoted on a stock exchange. Clearing banks and private loans are more important sources of credit than the stock market. In Britain, 64% of all credit raised by tenant farms and 40% by owner-occupiers and landlords is owed to the clearing banks and 17% and 38% respectively to private sources (Bosanquet 1968). The difficulties experienced by farmers in raising credit have led to the development of mortgage companies and co-operative banks dealing specifically with agricultural business. In many underdeveloped countries the moneylender is the only source of finance in remote rural areas, a position which he is quick to exploit. In Ghana, Pedler records that 'Whereas a bank might charge 6 per cent per annum for a loan against which the lender has deposited title deeds to property, stocks or shares, or even documents of title to goods, the money lender against good security charges from 12 per cent to 18 per cent and the money lender who lends against the mere promise to pay charges usually at 50 per cent per annum' (1955, 194). In advanced economies the spatial network of credit sources is usually sufficient for there to be few problems of access for the farmer seeking credit. This is not necessarily the case in underdeveloped countries where banks are fewer and distances between them greater. Where farmers have to claim grants from a central fund it may be anticipated that their propensity to do so will depend upon their distance from it, but in developed countries information on financial matters diffuses rapidly and is probably so effective that factors other than location, such as the attitude of the farmer himself, are more significant determinants of the receipt of government aid. In underdeveloped countries this is more problematic as information diffuses much more slowly and aid does not always reach the location intended.

In Britain Fixed Capital investment (land purchase, permanent buildings) in agriculture amounts to between £5500m. and £6500m., depending on the value placed on land (Bosanquet 1968). After maintenance expenditure is deducted, the net annual return on this investment is between 1·65% and 2·00%. Returns to Working Capital (seed, livestock, regular labour) are difficult to assess but an average annual return of 12–15% on Total Tenant's Capital[1] is usual (Nix 1968). However, this varies considerably by type and size of farm (see table 5.1). Many farmers over-invest in working capital in search of technical satisfaction, assuming that technically up-to-date firms can best withstand economic fluctuations and release manpower either for leisure pursuits or managerial study (Black 1965). Likewise, it is generally recognized that the average rate of return on capital in peasant farming economies is low (3–5% p.a.), while the interest rates on loans are high (10–15% p.a.); yet moneylenders lend money and farmers repay loans. In examining this paradox in south-east Asia, Long notes that although the debts of poorer farmers are higher in relation to their total wealth than those of richer farmers 'it should be recognized that many small farmers have little debt. In India 31% of farmers and in Thailand 33% had no debts at all. And those with little or no debt were primarily small farmers; for example, the average land holding in Thailand of those with no debt was only 57% as large as the holdings with debt and the average income of debt-free farmers was only 61% as great as that of indebted farmers' (1968, 1005). This would seem to imply that returns vary considerably between farms and that in a majority of cases it is only the large creditworthy businesses that can obtain credit. Moreover, farmers on the larger holdings are often the community's innovators who grow most of the cash crops, the only crops that realize sufficient profit for the repayment of loans with such high interest rates.

In developed economies much agricultural capital is tied up in long-term investments. The allocation of capital resources to alternative

1. This includes the tenant's contribution to fixed capital investments and machinery, as well as Working Capital.

long-term investments is therefore a major management decision. In an industry renowned for its uncertain physical and economic environments the assessment of rates of return and degrees of risk between alternative investments assumes special significance. Once capital has been invested in a particular enterprise, capital usually being available only in small quantities to small firms, the number of alternative land-use strategies that can be implemented at some time in the future if the

TABLE 5.1 *Average annual return on tenant's capital by type and size of farm*

Farm-type group	*Average area (ha)*	*Return on tenant's capital (%)*[1]
Mainly milk		
Under 50 ha	36 (89 a.)	11
50–80 ha	63 (155 a.)	15
Over 80 ha	111 (274 a.)	16
Mainly arable		
Under 100 ha	61 (151 a.)	7
100–180 ha	129 (318 a.)	20
Over 180 ha	287 (701 a.)	17
Milk and arable		
Under 100 ha	65 (160 a.)	13
100–180 ha	141 (348 a.)	17
Over 180 ha	253 (625 a.)	18
Sheep/cattle and arable		
Under 100 ha	65 (160 a.)	6
Over 100 ha	170 (420 a.)	16
Mainly pigs/poultry	47 (116 a.)	11
Pigs/poultry and arable	89 (220 a.)	8
Pigs/poultry, sheep/cattle, and arable	61 (151 a.)	9
Pigs/poultry, milk, and arable	117 (289 a.)	18
Intensive arable		
Fruit	40 (99 a.)	21
Vegetables	87 (215 a.)	22

1. Return on Tenant's Capital = Management and Investment Income as a percentage of Total Tenant's Capital. In calculating management and investment income (i.e. gross output less total inputs), the value of unpaid manual labour and an estimated rent on owner-occupied land are included in inputs, and land-ownership expenses, interest payments, and paid management are excluded). Based on Nix 1968, 102.

first strategy fails is greatly reduced. This is especially the case with perennial crop monocultures, as on rubber or banana plantations, since there is no income from these crops until several years after the major capital investments have been made.

Land

Despite the immobility of land as such its use is very mobile or variable, and the use of agricultural land much more so than most urban land. With the exception of perennial tree crops, permanent grass, rough grazings and the continuous sowing of the same arable crop, for example, padi rice in overcrowded areas of southern Asia or barley in parts of Britain, the use made of any particular plot of agricultural land changes frequently. Leys are often kept for three years or less and most arable crops are annuals. On fertile tropical land, especially where irrigation is practised, several crops can be taken in a single year. In southern Japan two rice crops a year are normal. Catch crops may also be taken in temperate zones. In Britain some farmers sow kale immediately after harvesting a cereal crop and then plough the land again in the spring after the kale has been consumed so as to sow a crop of spring barley. Indeed, until recently the use to which arable land has been put in Britain has rarely been the same in successive years, although technology has now largely removed the need for fixed rotations. This flexibility of use over short-time periods contrasts strongly with urban and industrial, and even forest, uses of land. Inevitably the inherent chemical and physical properties of the land vary spatially and impose varying limits on the agricultural use of the land, although actual use will be dependent upon technology, profit, and cultural constraints (see chapter 4). In particular its price, tenure systems and the demands being made on it for food production by a dependent population require comment.

In recent years the price of agricultural land in advanced economies has risen dramatically. In the United States land prices have risen four times faster than farm incomes since 1952, and in the state of Connecticut farm incomes fell by 10% between 1955 and 1964, while land prices rose by 59% (Chryst 1965). This price rise can be explained by interests outside agriculture competing on the land market and by farmers buying land at inflated prices in order to gain economies of scale (see chapter 6) by adding land to their existing units. Most British farm incomes do not justify the current average land price (£570 per hectare or £230 per acre approximately in 1969) at interest rates on borrowed capital of 10%. Those farmers who pay inflated prices need to obtain high returns per hectare to justify the cost, and this has been a major contributory factor in the intensification of farming systems in recent years. Intensification on arable farms has usually been achieved by the addition of extra cereal crops to existing rotations with the result that cereal production is now both widespread in the west of England and continuous in parts of the Midlands and south where mixed or even pastoral farming systems used to dominate.

Economic reasons, sometimes unrelated to agriculture, partly explain these high prices. In Britain land purchase is viewed as a hedge against inflation. Special treatment is afforded holders of agricultural land when estate duty has to be paid and there is always the possibility of urban development. Land worth £198 per hectare for agricultural purposes was being sold for £2,740 per hectare for housing development in 1950 and the same land for £9,880 per hectare in 1964 (Denman 1965). This speculative aspect of the land market is largely responsible for the higher land prices in the West Midlands (£573±18/hectare) and south-east England (£549±20/hectare) than in eastern England (£483±15/hectare) where much of the best arable land exists (Clery and Wood 1965). In the United States investment in farmland for recreational use is noticeably affecting prices, particularly those of poorer agricultural areas. Non-economic reasons also play their part. Many farmers who bought or inherited their land when prices were much lower fail to cost their land at current prices when evaluating the profitability of their businesses, and this is one reason why they are reluctant to cash their land assets and invest the proceeds in other more rewarding sectors of the economy. A survey of landowners records that the primary motive for 20% of those purchasing land is the desire to have the farm as a home, 18% do so for a sense of inheritance, 16% a feeling of social responsibility, and 9% a love of the land and ownership for ownership's sake (Denman 1965).

Land-tenure systems can have far-reaching effects on the land-use pattern. Numerous tenure systems exist and although classified here into three basic types, they can be more realistically considered as a continuum based on the degree of independence the farmer has to make what farming decisions he likes.

OWNERSHIP

The economic advantages of ownership include freedom to choose a system of production, returns for all improvements, increasing capital gains as land prices rise and, with the latter, increasing capacity to borrow. It also carries many social distinctions and as a tenure system is of increasing importance in the United States and Britain.

COMMUNAL OWNERSHIP

In this case the land is owned by the community or by the state on behalf of the people, although individuals may be responsible for working a number of plots. The village, as a corporate body, may hold land on behalf of its constituent families and allocate land according to their needs or rights. In the Soviet Union the state farm (*sovkhoz*) is owned and managed directly by the state, while on the collective farm (*kolkhoz*) the land is held collectively by the farm's workers. The latter is run partly in accordance with state directives, usually concerning the quotas of land to be allocated to particular crops.

TENANCY

Tenancies can lead to a form of business partnership between landlord and tenant, the landlord providing all or part of the capital for major farm improvements and the tenant paying for these through higher rents. A share-cropper hands over to the landlord a proportion of his output in kind as rent. In effect he is a kind of labourer as the landlord often provides the equipment, accommodation and even livestock, but unlike the wage-earner he shares his risks directly with the landlord. In advanced economies, independent bodies, such as rent tribunals, have been set up to protect the interests of both parties; but in many under-developed countries landlords can extract extortionate rents without the danger of recourse to an independent body.

The effects of land tenure on farm organization are numerous and complex and only by way of example can some of the more significant issues concerning tenancies in developed countries be referred to here. These are the lease itself, rent/farm income/land price relationships and farm size and fragmentation.

The lease

Two particular aspects of any lease are important. Firstly, the security of tenure it offers, and secondly, the extent to which the tenant is restricted in his farming operations by clauses in the lease. Security of tenure depends largely on the length of the lease. The longer the period the greater the security and the more willing the farmer is to invest his money, perhaps in conjunction with the landlord, in farm improvements. Insecurity tempts the farmer to over-crop and exhaust the land in an effort to obtain as large a profit as possible while he remains in possession. Leases vary considerably in length but tend to be shorter when land is in demand. In Ireland, for example, a tenure system known as 'conacre', by which the land is only let for eleven months at a time, is regularly practised. In some situations, however, little difference can be detected in the management policies of owner-occupiers and tenant-farmers. This often reflects the greater security of tenure in practice than is suggested by the short length of many leases, as well as the compensation rights accorded to farmers on dispossession for any improvements they have made to their holdings. In some leases the tenant is restricted in the farming systems he may employ to prevent the land being rundown. In Britain the implementation of clauses that stipulate particular cropping rotations for particular fields has been largely abandoned.

Rent/farm income/land price relationships

It has often been assumed that farm rents are directly related to farm incomes, and land prices to rents and incomes, but in reality the situation is not as simple as this. Although rents are higher where farm incomes are greater, they generally represent a smaller proportion of total income than when farm incomes are low. These relationships vary over time (table 5.2). For example, despite lagging behind, farmland prices

TABLE 5.2 *United Kingdom (excluding Southern Ireland throughout) residual incomes of land (i.e. economic rents) and land values*

	Economic rent net of maintenance receivable by owner of land (i.e. net of tithes, rates, etc.) £/ha/year	*Economic rent capitalized at current rate of interest on Government bonds* £/ha	*Actual price of land* £/ha	*Ratio of price to capitalized economic rent*
1867–73	1·58	48·90	111·90	2·28
1874–8	1·11	35·00	128·50	3·65
1879–83	0·00	0·00	93·70	—
1884–96	0·25	6·60	61·70	7·13
1879–1910	0·57	10·00	50·10	2·51
1911–14	1·36	46·20	56·80	1·38
1923–9	1·58	35·30	68·30	1·96
1930–2	1·43	34·30	56·50	1·65
1933–5	2·86	93·70	63·20	0·67
1936–8	2·79	87·30	67·60	0·77
1947/8–1951/2	12·85	370·50	196·20	0·53
1952/3–1956/7	12·25	287·00	187·50	0·65
1957/8–1959/60	10·77	239·90	233·90	0·97
1960/1–1962/3	14·03	235·70	332·70	1·41
1963/4–1965/6	14·10	228·00	543·50	2·38

Based on Clark 1969, 19.

have risen roughly in accordance with farm incomes in recent years in the United Kingdom. Rents, on the other hand, have not increased at nearly the same rate, and despite recent rent increases tenant-farmers still pay less for the privilege of farming and retain greater capital liquidity than owner-occupiers. This disparity only serves to emphasize the importance of non-economic motives in the purchase of agricultural land, and the lag on the part of landlords in adjusting to rising land price or farm incomes. In some cases, however, landlords raise rents prospectively. Rutherford (1966) records that landlords in Penang Province, Malaysia, raised the rents of all farmers when they heard that a few innovators had adopted a new variety of rice permitting double cropping.

Farm size and fragmentation

A land-tenure system based on tenancies is unlikely to lead to a successive reduction in farm size or an increase in the fragmentation of holdings unless the landowner can make a greater net profit from the rents of several small farms than from a single large one. The owner-occupier can divide his holding as he pleases and if local custom favours gavelkind (division of wealth between all sons or children) rather than primogeniture (eldest son or child receives all) average farm size can decline rapidly. Farm size is automatically reduced by land redistribution schemes which take land from the large landowners and divide it out among the landless and small peasant producers. Land reform is costly and has profound social consequences, but what is socially just may not be economically efficient or politically tenable, existing landlords often retaining holdings of above average size.

Population pressure

Many of the problems of underemployment, subdivision of holdings, lack of capital, and indebtedness occur together in underdeveloped countries, particularly in those frequently described as suffering from over-population, that is the condition in which any further increase in population will result in a decrease in the level of real income. Malthus first publicized the spectre of over-population in his book *Principles of Population* (1798). In this he claimed that while population increase occurred geometrically agricultural output could only be raised arithmetically. Despite concern over the rapid increase in the world's population in recent decades, substantiation of this claim, taking the world as a whole, is difficult, although certain countries (see table 5.3), particularly Iraq, Syria, and Burma, demonstrate a disturbing disparity between population growth and agricultural output. Nevertheless national figures also hide substantial regional differences within particular countries, and these persist despite international aid schemes, largely because of the existing social and administrative structures and poor communications of these regions. Neo-Malthusian concepts also emphasize that although recent technological developments have led to large increases in food output, over-cropping may result in the long-term degradation of land resources, and this issue is given scant attention by many economists concerned with raising output per man. Over-population is most serious in countries with high birth-rates and whose limited agricultural land resources are already intensively used, as in Egypt.

Other economists (Boserup 1965; Clark 1967, Ch. 4) have presented evidence which suggests that the problem of over-population has been exaggerated and is less widespread than previously implied, occurring only locally, as in the densely populated region of south-east Nigeria. Boserup (1965) attacks the Malthusian concept that implies that food

TABLE 5.3 *Growth-rates per cent per year – agricultural production and population*

	Agricultural Production 1934–8/1959–60	1953–4/1959–60	*Population* 1936–60
Israel		9·8	7·0
Guatemala		4·6	2·6
Iran		3·3	0·9
Panama		3·2	2·7
Turkey	3·2		2·2
Columbia	3·0		2·2
Thailand	3·0		2·6
Venezuela		2·7	2·9
Cuba	2·3		2·0
Ceylon	2·3		2·3
Phillipines	2·2		2·6
South Africa	2·2		2·1
Honduras		2·1	2·6
Tunisia	2·1		2·0
Peru	2·0		2·1
Malaya	2·0		2·4
Taiwan	1·7		2·9
Morocco	1·7		2·1
Egypt	1·6		2·1
India	1·3		1·5
Chile	1·2		1·8
Indonesia	1·2		1·4
South Korea	0·9		2·1
Pakistan	0·7		1·5
Algeria	0·5		1·7
Ethiopia		0·4	1·6
Burma	–0·2		1·2
Syria		–0·2	2·6
Iraq		–0·3	2·7

Rates computed from FAO Index Numbers of Agricultural Production.
Source: Clark and Haswell 1966, 73.

output is essentially inelastic and determines population growth. She suggests that this concept should be reversed and that increasing population pressure should be viewed as a stimulus for technological change in agriculture which will then provide the increased food necessary for the enlarged population. As evidence for this she cites the substitution of sedentary cultivation for shifting cultivation systems and a reduction in the length of fallow period in the crop rotation as population densities increase. To what extent this argument, largely derived from under-populated areas, could be applied to areas of padi farming in Monsoon Asia is open to question. However, there is little doubt that in Latin America, for example, population pressure could be reduced by the intensification of farming methods on the large estates.

Population density affects land-use patterns considerably and Boserup (1965) has gone so far as to claim that 'it is unrealistic to regard agricultural cultivation systems as adaptations to different natural conditions, and that cultivation systems can be more plausibly explained as a result of differences in population density' (p. 117). This assertion again needs testing in areas with contrasting population densities or varying densities associated with changes in agricultural practice. Few attempts have yet been made to do so, and considerable difficulties remain in inferring the relationships involved.

6 Scale of production

A more effective use of fixed resources can frequently be obtained by increasing the scale of business activity, and this is particularly true in an industry such as agriculture which contains many small production units. It is therefore in the farmer's financial interest to determine the scale of activity which will maximize his profits, as it may benefit government if an industry of sufficient size to make efficient use of national resources is established.

Two kinds of scale economy are generally recognized, that endogenous to the firm (internal economies of scale) and that exogenous to the firm (external economies of scale). This distinction often depends on whether the viewpoint of the firm or the industry is adopted. For example, an expansion in poultry production may result in government's setting up an advisory service specifically for poultry producers. This service may increase the efficiency of poultry production and while acting as an external economy to the individual firm effects an internal saving to the industry as a whole. If poultry production were regionally concentrated the government might also invest in a new regional infrastructure providing further economies and encouraging even greater concentration. This concentration could have a multiplier effect, creating a large, consistent demand for barley, enabling local barley producers to specialize. Barley farmers would then obtain internal economies of scale from specialization as a result of external economies derived from the demands of the poultry producers.

External economies have received little attention from agricultural geographers, but three examples are sufficient to indicate their significance for the spatial organization of agricultural patterns. Firstly, when discussing possible explanations for the increasing concentration in hop production in nineteenth-century Kent, Harvey (1963) examined the external economies derived by the hop producers from their localized distribution. He suggested that the pool of skilled labour, the special sources of credit and marketing channels available to producers in mid-Kent created sufficient economies to offset the high farm rents attributed to the concentration of this profitable industry (see pp. 132–34 below). Secondly, the Milk Marketing Board of England and Wales provides external economies for certain producers by its use of standard collection charges for milk (see below, pp. 96–97). Thirdly, the expansion of horticulture in the Evesham district has been encouraged by supplies of local skilled labour and the development of a specialized local market for fruit and vegetables, so that today the producing areas extend far beyond those that were traditionally regarded as being physically suited to the industry.

It is, however, with the farm's internal economies of scale that most geographers and agricultural economists have been concerned, and the remainder of the chapter examines this question.

It is extremely difficult to be specific about the economies attributable to an increase in scale of production, for as Heady (1956) has pointed out, scale and productivity are interdependent. Unless increases in total output can be related to increases in inputs in fixed proportions, true measures of scale relationships cannot be recorded. In reality, an increase in scale of production often leads to a change in the relative significance of each input, and hence their optimum combination, violating Heady's limiting assumption. Consequently, where data are available scale economies are discussed in terms of the single most limiting factor of production or, more often, simply in terms of farm size.

Farm size

Traditionally, land area has been considered as the primary criterion of farm size, probably as this fact is recorded in most bodies of agricultural statistics. Thus in Britain, for example, farms of less than 20 or 30 hectares (50–75 acres) are considered small and those over 120 hectares (300 acres) large. Yet land represents only one farm resource and it is the manner in which it is combined with other resources that determines the level of farm output and the size of business. Inputs, outputs and net returns per unit area differ substantially between enterprises and in some cases, for example, intensive pig, poultry and glasshouse production, require very limited land resources. Alternatively, standard labour inputs by enterprise have been employed by economists and geographers (Coppock 1965*a*) where the data exist (see pp. 109–11). Knowing the labour input required per hectare by enterprise on the average farm and the enterprise areas on each farm, total labour inputs, as a measure of business activity, can be assessed. Standard enterprise outputs by value have also been used in a similar manner (Jackson *et al.* 1963).

Although area is not the most limiting factor of production in many agricultural economies and is an unreliable measure of farm business size, it does bear some relationship to farm income. Its general availability (as opposed to farm output or labour input data) means that its use as a measure of scale is adopted here.

The size of the 'minimum farm', the smallest possible unit that will support a farmer and his family, depends entirely on the social and economic context in which the study is being made. The differing intensity of production systems, variation in economic circumstance, and what is acceptable to particular communities and individuals make international comparisons of little value. Under intensive, subsistence rice cultivation conditions one hectare may suffice. In Britain the Wise Report on Smallholdings (1967) stated that an income of £1,200 per annum was the minimum desirable, and in most circumstances this meant a farm of at least 30 hectares, while in the drier parts of the Great Plains farms of 200 hectares are considered small. The concept of farm size has varied temporally as well as spatially as yields have risen and mechanization has been introduced.

The small farm

The basic management problem of limited acreage is the restriction it places on the farmer's choice of farming system. In order to realize the minimum necessary income only those enterprises producing a high net return per acre can be considered. On many farms in Britain this means either dairying, the growing of cash root crops or the supplementation of

TABLE 6.1 *Inputs and outputs on small and large farms in eastern England*

	Small farms	*Large farms*
Average size (ha)	26 (64 a.)	103 (254 a.)
Farm income/ha	£47·70	£27·40
Farm income/ha *less* farmer's labour	£27·40	£19·80
Gross output/ha	£192·70	£153·70
Net output/ha	£134·40	£118·80
Costs/ha as % of gross output/ha	30·25%	22·35%
Net output/£100 labour	£213	£275
Net output/£100 labour and machinery	£170	£212

Based on Sturrock 1965, 8.

an existing system by more intensive methods such as housed livestock production. These systems of production require extra capital normally not readily available to the small farmer who is often less efficient in his use of capital than farmers with larger units.

Economies of scale are difficult to achieve because of the indivisibility of many inputs. Although gross outputs per hectare are generally higher on small than large farms, costs consume a greater proportion of the gross return (table 6.1). High costs dictate specialization and the need to obtain all possible internal economies, but this may in turn lead to an inflexible farming system in which certain advantages derived from a mixed system of production are lost. For example, seasonal variations in labour demand may be increased and in marginal arable areas continuous cropping may lead to a deterioration in soil conditions.

Surveys indicate that there are certain sociological characteristics more attributable to farmers of smallholdings than of large farms. For example, the average age of the farmer tends to be higher than the mean (see Urquhart 1965; Wise Report 1967). This is particularly significant as middle-aged farmers have less opportunity to obtain capital for expansion and are less willing to innovate than their more youthful counterparts. Furthermore, beyond the desire to remain independent, the aspirations of small farmers are limited (Thorns 1968), although it is argued that there is little concrete evidence to support the assumption that small farmers are inefficient out of choice or like their low incomes and insecurity (Nalson 1968). As a group, their conservative outlook often leads to the rejection of new management methods, but this is not surprising as many live in a social environment that equates work with manual labour. In addition, although many small farmers claim to have little time to spare in the search for market or technical information, their limited use of free advisory services, probably the most effective means of obtaining information quickly, and at the source of the problem, the farm, leads to concern.

Brief comment should also be made on two other important types of farm, often associated, sometimes incorrectly, with the small farm – the family and part-time farm. The term family farm is a sociological one, referring to farms dependent upon family labour. The family farm still predominates in the underdeveloped world and despite the trend towards larger holdings in the developed world it is of increasing significance where hired labour is rapidly leaving agriculture. The high incomes of these large, mechanized units contrast with the traditional poverty and near subsistence economy generally associated with the family farm. Yet despite mechanization, a major long-term management problem still faces family farms, that of the cycle of family development (Nalson 1968). Where the family represents the sole available work force, the size of that work force depends on the number of family members of working age. If farm size is fixed and the substitution of capital for labour is impracticable, then the level of efficiency of labour use and the kind of production system employed will vary according to the number of family members of working age. Only within those societies that have complete mobility of labour or a communal ownership of land can these changing labour circumstances be fully taken into account.

A part-time farm is usually defined as a unit that fails to provide full employment for one man. In the United Kingdom farms with labour inputs of less than 275 Standard Man Days[1] per year are considered part-time holdings and in June 1955 there were 180,000 such holdings

[1] 1 SMD = an 8-hour working day.

(including some multiple-units) in England and Wales or 48% of all farms (Ashton and Cracknell 1960–1). By this definition, the under-employment of labour on many farms in the underdeveloped world would result in many more holdings being classified as part-time units. It is, however, the increase in number of part-time units in industrialized nations that has received most attention. Generally, the full-time

TABLE 6.2 *Part-time farmers in south-east England classified by the socio-economic group of their other occupations*

Socio-economic group	*Description of occupation*	*Proportion of part-time farmers* (%)
2	Agricultural workers	8
3	Higher professional, administrative or managerial	35
4	Lower professional, administrative or managerial	23
5	Owners or managers of wholesale or retail businesses	11
6	Clerical workers	5
7	Shop assistants	
8	Personal service	
9	Foreman	18
10	Skilled workers	
11	Semi-skilled workers	
12	Unskilled workers	

Taken from Gasson 1966*b*, 21.

farmer does not become a part-time farmer, but if he has no son to succeed him he may enter a period of voluntary semi-retirement. He either struggles to keep a non-viable unit going or he leaves the industry altogether. He rarely takes part-time employment off the farm, although other members of his family may be forced to do so. It is largely people from outside the industry, often professional people (see table 6.2), who buy farms within commuting distance of their urban occupations. These farms are not always small in acreage but simple systems of production requiring low labour and management inputs are favoured (see table 6.3). Non-profit making motives, such as the farm as a home, leisure and prestige, assume greater significance than on full-time farms (Gasson 1966), and the farm-owner often becomes the occupier but not the operator. There is less urgency to adapt to changing economic conditions or to be efficient, as income from the urban occupation reduces dependence on the farm's profits and a further obstacle to farm structure reform is created.

TABLE 6.3 *Number of enterprises per full-time and part-time farm from sample survey in south-east England, 1965*

Number of enterprises	*Number of farms*		*Proportion of farms* (%)	
	Full-time	*Part-time*	*Full-time*	*Part-time*
1	11	23	15	28
2	8	24	11	30
3	20	15	27	19
4	12	12	17	15
5	12	6	17	7
6	4	1	5	1
7	5	—	7	—
8	1	—	1	—
All farms	73	81	100	100

Part-time farms differ significantly from full-time farms at the 1% level
Taken from Gasson 1966*b*, 39.

Intensification and co-operation

If extra land cannot be bought or rented the small farmer can only increase his profits either by raising output from his existing holding or by co-operating with other farmers in order to reduce costs. In Britain, increases in output in recent years have been obtained either by more intensive cropping or stocking, or from factory farming. In the latter case, the animals are kept permanently in buildings and make no demands directly on land resources. Intensification requires capital investment in machinery and buildings and this has encouraged some farms to seek benefits from horizontal integration.

The degree of co-operation in agriculture varies according to local or national traditions. Complete co-operation based on the communal ownership of land, as on the *moshavim* of Israel, contrasts strongly with the traditional concept of the independent British smallholder. In between these extremes are rural communities in which co-operatives are encouraged and financed by the state, for example, Denmark and Finland, where such services as agricultural education and product quality control are provided. This spirit is often reflected in the extent and variety of co-operation between individual farmers, as table 6.4 shows for a random sample of 200 farms in south-west Finland (Honkala 1969). In Britain, co-operation is less well established, although in recent years the indivisibility of certain major capital expenditures, for example, combine harvesters and grain silos, has led to an increase in the number of machinery syndicates and more joint use of storage facilities. Similarly, in order to improve their bargaining position, farmers have formed co-operative marketing groups, particularly for vegetables, and have attempted to reduce costs by bulk buying. The deliberate exchange of feed and stock between farms is also increasing, but the complete integration of businesses and their managements is infrequently found. In Britain the small farmers' desire for independence and the administrative difficulties of running a co-operative based on numerous small units has resulted in the medium-sized rather than the small farmer being actively involved in co-operation.

TABLE 6.4 *Varieties of co-operation and participation in them in south-west Finland*

Variety of co-operation	*Number of cases*
Labour exchange	103
Machinery exchange	54
Machinery and labour exchange	47
Co-ownership of machinery	90
Co-operative purchases	9
Joint pasture	6
Joint forest	9
Total	318

Taken from Honkala 1969.

The large farm

Large farms may simply have extensive land resources, as for example the Italian *latifundia*, or be a vertically integrated business, such as the plantation or the intensive beef-lot. A large acreage permits the efficient utilization of capital and labour, and the low yields associated with extensive production are in many cases the result of a management policy aimed at the best use of poor land rather any inherent management weakness.

Although the traditional concept of the plantation has been challenged (Gregor 1965; Jackson 1969), the plantation remains a unit of large-scale production in relation to its peasant farm competitors. It gains certain economies of scale, in relation to the peasant farmer, from being close to sources of technological information, employing a skilled labour force, having an efficient processing plant and from established market outlets. Today, however, social and political forces have undermined these economic advantages and where the peasant farmer is able to discount the cost of his own labour he is in a strong marketing position.

The number of farmers in advanced economies who are developing a vertically integrated business, that is who manage their marketing facilities as well as their production processes[1] is increasing fast. This

[1] Such vertically integrated businesses are commonly known as agribusinesses.

development can be expected to affect substantially the future location of many agricultural activities, particularly as it is the large producer selling considerable quantities of a standard quality product directly to the retailer who gains the greatest economies from vertical integration. The enhanced marketing expertise required of these farmers will lead to farming systems more sensitive to changes in the location and structure of market facilities than those of other farmers. By trading directly with the retailer, much of the price uncertainty is removed for the farmer and profit margins can be increased. This favourable position will only remain, however, as long as other farmers are obliged to sell their produce, because of its limited quantity or uneven quality, in

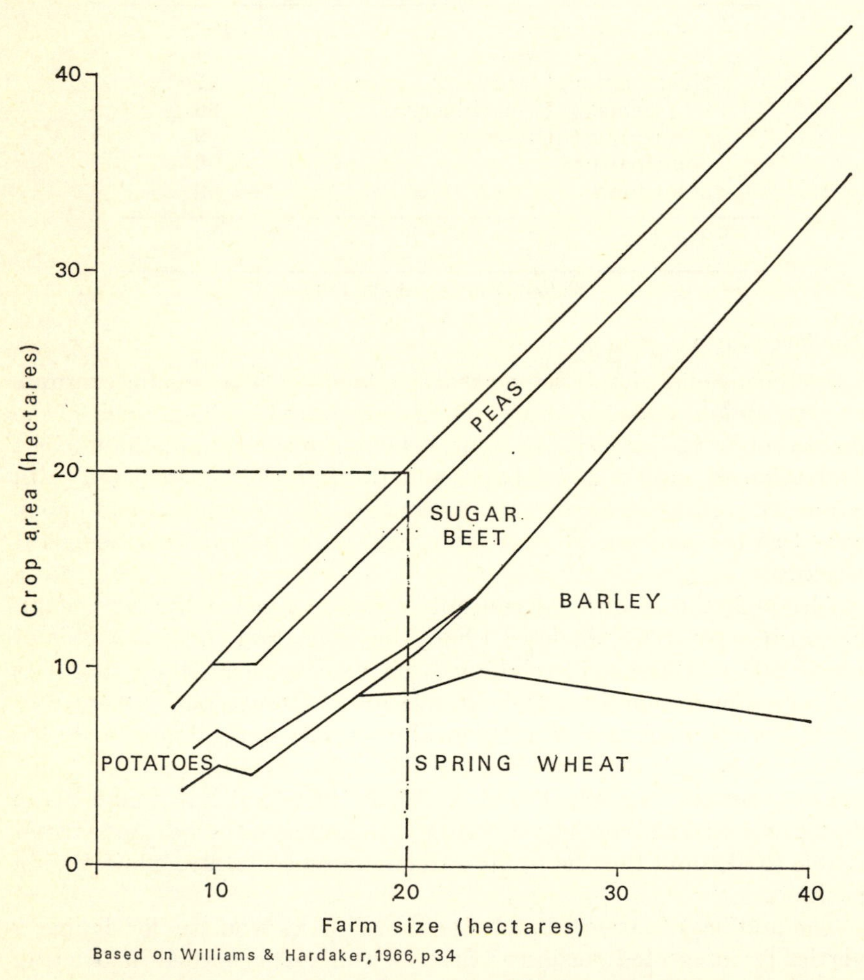

6.1 *Optimum cropping mixes by size of farm: an example from the English Fens.*

traditional markets, and more than one retail outlet is available to the farmer.

It is possible to gain an insight into future developments from a study of the highly competitive and vertically integrated broiler industry of North America (Burbee *et al.* 1965). As a result of their large-scale production, farms in this industry have been forced to assess their transport costs more critically than hitherto, although these represent only a small proportion of total costs. Consequently the location of production, at present often at one of the sources of feed (the original farm), may be gradually replaced by a more market-oriented location.

Large farms are not necessarily socially or economically desirable. At a time of food shortage it would be preferable to have an industry consisting of small farms with a high level of output per acre. This could lead to a conflict between public and private interests if farm size increase, with associated economies of scale, were prohibited by the state. Furthermore, large units cannot always be justified on economic grounds in underdeveloped economies. Where employment possibilities do not exist outside agriculture, an extensive use of labour on large holdings will lead to national diseconomies as marginal returns to the national labour force decline and physical output falls. Economic efficiency is not found on many Latin American estates (Feder 1965) as the estate owners deliberately maintain labour underemployment as a means of keeping down wages and making labour obedient, and the limited profits that are made are not reinvested in agriculture.

The search for an optimum size of farm is as unrewarding as attempts to define the size of the minimum farm, and for the same reasons. What is economically optimum may be socially undesirable. For example, one might be tempted to say that 70 cows on 36 hectares (89 acres) would be close to the optimum size for an up-to-date one-man dairy farm in Britain today – provided the farmer had few thoughts for holidays or other leisure pursuits. More often economists have been content to accept the land resources of a particular farm as fixed, and have attempted to maximize potential profits by budgeting or linear programming techniques (see fig. 6.1).

7 Marketing and supply

Despite the emergence in Europe and North America of hobby-farmers and part-time farmers who have interests elsewhere, and despite the tendency of many farmers to develop somewhat irrational preferences for certain kinds of enterprise, the market is undoubtedly one of the most potent factors in agricultural production. We may observe this at any period of changing prices or of improved technical efficiency making possible either lower prices or greater profitability. The remarkable spread of feed barley in Western Europe in the past fifteen years, of hybrid corn in the United States, or of coffee, cocoa and rubber in tropical countries all witness the power of the market and the desire of producers to take advantage wherever gains are to be made. Unfortunately for the farmers, their response to an advantageous market is eventually to offer more of a commodity than demand requires. Because of the large number of small firms in farming there is a constant tendency to over-supply with in consequence poor prices, small profits and low wages. Agricultural prosperity has most frequently been associated with periods of rising demand. The growth of the great food markets in the industrial cities of North America and Europe, where in some cases they have been encouraged by protectionist policies limiting food imports, has brought a greater prosperity to farming than it has ever previously known. However, a near-stable population situation with food wants mostly satisfied and no outlet overseas can lead to overproduction and a fall in prices unless they are artificially maintained by government intervention.

Apart from problems such as dumping, the agricultural commodity markets are generally controlled by buyers rather than sellers, and the only major ways in which farmers can protect their interests are by forming groups of sellers, by creating an agency to act on their behalf, or by persuading government to interfere with the operations of the market. However, farmers can influence the market by obtaining better market information and by storing their products on the farm until prices are satisfactory. The growth of direct dealing and contract systems of selling produce also represent attempts to avoid the risks of traditional marketing and to improve the financial return for the farmer. Nevertheless, even in direct dealing with merchants, with canning or freezing firms, or with supermarket chains the farmers are generally in a weak bargaining position because their numbers are so much greater than the number of potential buyers.

These weaknesses have important locational differences. Most countries differ in their marketing systems, more especially in government-controlled constraints or inducements, in the size and kinds of market offered and in the development of overseas trade. Thus some countries have marketing boards which strictly control product prices, others have subsidies or guarantees which may vary according to the state of the market. In Britain the chief inducement to dairy farming lies in the liquid milk market and a system of controlled milk prices. In the E.E.C. the price guarantees are on butter and cheese, and dairy production on the small farm tends to look more to a butter and cheese market than a milk market. Marketing organization methods differ not only between countries, but also between commodities. Some commodities lend themselves much more to one method than another. For example, group trading by farmers is more common among vegetable and fruit growers. Members of the group contract to produce whatever their marketing organization tells them can be sold at acceptable prices. Grain, on the other hand, is generally handled by merchants and is much more subject to the vagaries of international trade than are fruit and vegetables despite the recent spectacular growth in world trading of these commodities. In Great Britain grain is bought either through the London Corn Exchange where relatively small parcels are sold, or through the Baltic, i.e. the Baltic Mercantile and Shipping Exchange, where large parcels, even whole cargoes, are sold, or through the Liverpool Corn Exchange or the Pacific Limited at Hull. Australian and Canadian wheat is sold by wheat boards or their agents by a price arrangement which ensures that when the wheat arrives its price will approximate to spot value. No such arrangement exists for wheat purchases from the United States, where the price risks may still be considerable (Rees and Wiseman 1969). This difference is in effect a factor in the grain trade and indirectly of world wheat location. Dealings in London dominate the world grain trade, and on some occasions almost the entire world's exports of grain have been handled by British dealers. Organization for the marketing of perishable commodities such as liquid milk, has been carried out most successfully by the formation of co-operatives which in certain cases, especially in Denmark, have brought standardization and grading of produce to high levels. The development of brand names, for example, in dairy produce, frozen chicken and bacon has helped to ensure for some farmers some guarantee of continuing sales in certain markets.

The largest international markets in agricultural produce are in the advanced industrial countries. Thus the markets of, say, Birmingham and Glasgow have hinterlands extending to Argentina, Australia and

New Zealand. The United Kingdom depends on the efforts of farmers in these and many other countries, but they in turn depend on continuing to find markets in the United Kingdom until some alternative markets appear. Countries with mixed economies like Australia are less dependent and run less risk when one considers their economies as wholes than countries such as Senegal, New Zealand or Ireland whose exports are made up largely of agricultural produce and sometimes, as in the case of groundnuts from Senegal or cocoa from Ghana, depend mainly on one product. The inelasticity of the basic food markets of the 'developed world' has already received comment (p. 8). It may hinder expansion in crop production and may even lead to a situation of chronic overproduction. Where trading arrangements are carefully controlled they may encourage a production situation of great stability leading to the development of a constant or steadily increasing production area. On the whole the 'underdeveloped' or 'developing' world suffers from a trade dependence on industrial commodities, such as cotton and rubber, which are liable to suffer from considerable fluctuations in production and in world market prices. To suggest that for all their agricultural dependence countries like Ceylon and Kenya can emulate Denmark and Holland in developing overseas markets for their farm produce is to ignore the enormous difference in the nature of the produce offered and the importance of established trading links and agreements. Not even the development of government-controlled marketing boards, nor of production control conferences attended by representatives of the producer countries, has succeeded in preventing considerable price swings.

The apparent effectiveness of the market as a location factor is frequently extremely difficult to judge in relation to other factors. One may see apparent evidence of it in locational maps, one may seek the opinions of farmers or dealers, or attempt to measure its possible effectiveness by comparing prices in different markets and examining the effects of transfer costs as between one market and another. On a given farm the farmer may consider market factors as important, yet comparison between his enterprise pattern and that of another farmer may fail to reveal them. Only when a large number of farmers is compared is the fact apparent. Distance is important as this relates significantly to transport costs and the size of a given market centre. The greater the distance generally the greater the transport costs, although frequently this is offset by tapering effects, i.e. the reduction in costs per ton mile with increasing distance. Commodities which find their main markets at considerable distances away may be very sensitive to cost differences between one market and another, although frequently some of these commodities, such as wheat or maize, may be handled cheaply by modern bulk-cargo methods. Large markets in agricultural produce necessarily, however, depend on large areas and high transfer charges for commodities from the perimeters of those areas. These perimeterward locations may be expected to show greater variation in market choice than locations near a given market centre. They may also show more limitation in the range of enterprises considered remunerative. Differences in size of market with differences in radius of operations may also mean differences in the frequency of visits and this may relate to the commodities offered. Many crops need only a once a year market, but livestock may be bought and sold much more often, hence the greater persistence of small livestock markets in certain countries. Some producers prefer a particular enterprise simply because they like going to market, and enjoy either bargaining or the social aspect of the occasion.

Livestock and produce markets

A very large part, possibly most, of the world's agricultural produce is either consumed at home by the farmer and his dependents or is sold in local markets where the prices are fixed by bargaining or the goods exchanged by barter. The variety of exchange systems is considerable. It has been claimed that there are as many exchange systems as there are types of society (Belshaw 1965, 7). Primitive systems may involve simply the exchange of gifts, ceremonial exchange, trading partnerships, or 'silent' exchange. In most peasant societies where money is used as a medium of exchange there is some production for sale and some production for local subsistence. Often the same crops are involved in both. Frequently in such societies markets occur in the form of a hierarchy with local village markets serving immediate needs and urban markets serving as foci for the bigger farmers and dealers who may 'bulk up' their wares by numerous purchases in the smaller markets.

In the more commercialized societies, selling methods, transport and distribution are more varied and much more effective. Very little produce is sold directly to consumers, although producer-retailers farming on the urban fringe and selling farm-fresh eggs and butter operate in the suburbs of large cities. Distance is less of a problem and specialization more attractive to the farmer. In Western Europe, in addition to the larger livestock markets handling more than 100,000 head a year, local farmers' auction markets, handling less than 10,000 head a year, still exist. This need of local livestock markets continues because apart from high-quality meats, part of the demand can be met locally at minimal transport cost. The need also arises from the continued existence of very large numbers of small-scale livestock farmers offering only a few beasts at a time for sale or wishing themselves to purchase a few replacements for their herds. For such purposes local markets where the dealers and farmers attending are all well known to one another are preferred. In Great Britain most beef comes from the dairy herd (75% in 1955, Coppock, 1964*a*, 161), as does most veal in the E.E.C., and partly because the dairy herd is so widespread there is still an abundance of small slaughtering plants. These are being reduced in numbers in most countries as public abattoirs replace private slaughtering. This growing tendency to concentrate slaughtering in a few centres and the growth of larger herds has increased the competitive advantages of the larger markets. The trend is reinforced by the fact that the larger markets tend to receive more notification from sellers of the entry of livestock than smaller markets. They are thus in a better position to attract buyers by advising them in advance of what stock will be on offer (Weeks and Brayshaw 1966). If the current trend in fatstock sales towards an increasing use of the larger markets continues this will reduce an important if lesser part of the business at the smaller markets and may mean that for some the remaining business in store cattle is insufficient on its own for them to survive.

Putnam (1923, 19) has argued that great distances necessitate large slaughtering plants, hence the vast stockyards and abattoirs of Chicago and the development of large-scale meat packing in Britain with the growth in the amount of imported meat. However, the decline of meat exports from the United States has brought in turn a decline in the importance of Chicago (Cairncross 1966, 137), no doubt worsened by direct sales from farms to supermarket chains. Greater dependence in the United States on the home market has led to some decentralization of meat packing and the setting up of smaller slaughter-houses.

Perishables, that is fruit and vegetables, are increasingly being bought directly by dealers, sold to large packaging, canning, freezing or drying firms, or sold to large retail firms owning supermarket chains. In Great Britain most horticultural produce is still handled by the thirty large city wholesale markets. Only very small amounts are still sold in local auction markets. Frequently the price of vegetables is higher in retail shops near the areas of production than in shops in large cities, either because of the greater effectiveness of the city markets or because local retailers are unable to buy locally due to grower's preference for contracts with the larger buyers. Crops such as peas need to be frozen or processed quickly if they are to retain their quality, and speed of packaging or marketing is especially important for the soft fruits.

Direct sales of produce

In the industrial countries a great deal of agricultural produce is bought from farmers directly by dealers or produce boards. This is especially true of the major grain and root crops, but it is also becoming of some importance in livestock and fruit and vegetable sales. The dealer is normally better equipped than the farmer for the sale of farm produce to advantage and may store produce until a favourable moment for disposal. Many farmers attempt to obtain a large part of their short-term credit from dealers, either from those who buy their crops or from those who sell materials and equipment to them. For the grain grower who depends mainly on one major harvest a year and whose income may be subject to considerable risk, dealers' credit may be a vital factor in his organization and its availability or otherwise may determine his choice of enterprise. In many countries the development of co-operative organizations which act on the farmers' behalf such as the growers' associations in Britain or co-operative selling agencies in France, have altered the relations between farmers and dealers. Merchants in self defence are tending to group themselves together, as for example in cocoa purchase in West Africa, and have to integrate themselves into marketing schemes. Large farm businesses are frequently more able to find their own capital for short-term needs or are better able to obtain it from alternative sources than small concerns. In consequence they have always been less tied to particular merchants.

Direct purchase by manufacturers is of increasing importance in most countries. In the developing countries the need to export processed rather than raw materials, together with the need for a guaranteed supply of a given crop in terms of both quality and quantity, encouraged the early development of plantation systems established around a factory. This has certainly been a characteristic feature of sugar-cane production for some centuries and even in some cases of the production of cotton, rubber, tea and palm oil and kernels. The development of processing mills to serve peasant producers has been a feature mainly of this century and particularly of the last two decades. In Europe and North America the development over the past two decades of super-markets, sometimes organized in large groups themselves capable of direct purchase, together with the expansion in sales of packaged, prepared, ready-cooked, frozen, freeze-dried and canned foods have brought a revolution in food-crop marketing. The earliest dependence of farmers on sales directly to factories was among sugar-beet growers – in Great Britain the first factory was erected in 1912 – and among dairy farmers who sold their milk to creameries. The sugar-beet price in Britain, for example, is determined before the crop is sold, the crop being grown under contracts made each year between farmers and the British Sugar Corporation which has a statutory monopoly. The acreage and the prices are determined each year by the Government by a quota arrangement and this has tended to fossilize the production pattern (Coppock 1964*a*, 60).

Quick freezing requires extremely rapid processing after harvest and it is essential that crops are grown near the freezing plant. In some cases new plant varieties better adapted to the process concerned, or of superior habit to meet the size standards of the market or better adapted to machine packing, are being grown and together with the freezing factories are, in some cases, finding new locations (Coppock 1964*a*, 62). Increasingly potatoes are finding a market in canneries, or for potato crisps or frozen chips. Some 40% of the British crop is now packed or processed. Manufacturers establish long-term contracts with farmers, an advantage to small growers in an industry liable to suffer from slumps in price. Thus while the British potato glut of 1968 hit producers of early potatoes, excluded from the price guarantee scheme, and kept main crop prices down, contract growers were unaffected. However, the freezing and canning factories mostly draw on the local area. In Great Yarmouth, for example, one factory obtains nearly three-quarters of its potatoes for processing into frozen chips from within a 56-km radius. An interesting feature is that vegetable freezing is mostly done at eastern fishing ports where the quick-freezing plant treats both vegetables and fish, thereby reducing the seasonal swing in its use of machinery and labour.

Transport costs

Transport costs, including not only the costs of conveyance but also all the costs arising from the movement of goods,[1] have long been regarded by geographers as of special locational significance. In agricultural geography concern has been chiefly with the costs of transferring agricultural produce from the farm to the market or to the processing plant. The numerous models of agricultural location constructed to show the effects of transport costs have been based on a large number of assumptions and have normally ignored the fact that transportation is itself an industry with its own locational problems, with highly complex costing systems, and with a history of constantly improving 'productivity', so that in those countries with well-developed transport networks transport costs to market for farmers are on average frequently less than 5% of total costs, although varying considerably between different produce. In the Highlands of Scotland, for example, transport costs for livestock are no more than 2% of the purchase price in the crofting counties and only 1% in the remainder of the region (Stewart 1964). Most of the models have been based also on a profit maximizing assumption in which the farmer is presumed to have adequate access to relevant market information, together with an equally competent ability to compare the suitability of alternative markets and to act accordingly.

[1] Some economic geographers refer to total transport costs as 'transfer costs', but confusion may arise due to the use of the latter term by economists to refer to the cost of transferring a marginal unit of resource from one industry to another, that is, the notion of cost as relinquished alternatives.

Market attraction should not be explained by the transport factor alone, i.e. the cost of transport of crops and livestock to market. Size of market in relation to producer, prices offered, number and quality of alternative markets, accessibility and visiting frequency need are all important. Attempts have been made to assess market potential and relate it to transport costs. Harris tried to apply the gravity model to the problem and constructed maps of market potential and of transport cost to national markets in the United States (Harris 1954). Dunn tried to combine market potential and transport cost into a single index of optimal location (Dunn 1954). To do this he had to assume that a disadvantage in transport cost could be exactly offset by an advantage in market potential of equal proportion. There is, unfortunately, no basis for such an assumption (Isard 1960, 525), and the problem of relating transport costs in the form of a general index not only to market potential but also to differences in other costs such as labour and land, has not yet been resolved. The observation of land-use zones around peasant farming villages in Southern Europe or Africa and their obvious association with distance from their village centres should not lead one to regard the village as a market surrounded by zones of production of diminishing intensity outwards. Frequently, in such cases, the major factor is not the transport of crops but of the cultivator, together with the problems of frequency of visits and of the costs of conveying manure from the village to the fields (Morgan 1969).

Total transport costs include freight charges, insurance, customs charges (where required), transhipment costs, packaging and handling costs and storage charges. The size of market may be an important factor because a large market may encourage transport and handling innovation together with economies of scale. The example of the world trade in wheat is well known. Wheat is a bulky commodity for its value, and should be moderately expensive to transport. In practice wheat is easy to handle, and the development of huge international grain markets especially that of Great Britain which imports some 8 million tonnes[1] of wheat, maize and sorghum each year, has encouraged the development of special grain-carrying ships, the opening of new water-routes, such as that via Hudson Bay from the Canadian wheatlands and the construction of new railway systems. World wheat markets are highly competitive and tend to overlap although there is a regional pattern of market areas where one particular source of wheat is dominant. This pattern is partly determined by distance and partly by political arrangements (Lösch 1954, 420–7).

Transport costs are normally a function of weight and distance, and simple locational theory, such as that developed by Alfred Weber (1909), emphasized the role of weight of materials as a locational factor.

[1] i.e. metric tons. See appendix.

Weber developed the thesis that where there was a loss of weight of materials in the course of processing or manufacture, then the industry concerned was tied in location to its raw materials. Smith (1955) cited the locations of sugar-beet factories and milk-processing factories as examples, but noted that loss of weight had locational significance only when combined with large weights per operative 'for variations in transport costs are substantial enough to affect location only if weights are large'. Where their transport charges vary with distance, bulky or heavy crops may thus tend to be grown near markets. Coppock (1964*a*,

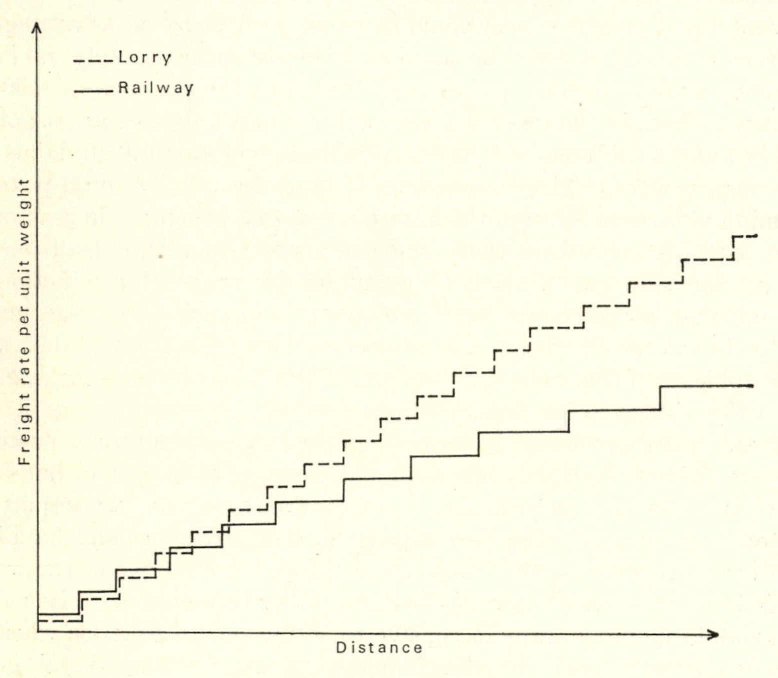

7.1 *Freight rate structures.*

Idealized freight rates showing stepping and taper. In general railway freight rates show a much greater tendency to taper than freight rates by road transport, which are often cheaper over short distances.

85) noted the relative importance of potatoes in districts near the main urban markets of England and Wales, where they occupied over 75% of the root acreage. In general terms, Shaw (1970) has found that this spatial pattern of potato production, although not optimal, is quite efficient when the haulage costs from producer to market are added to the production costs of potatoes in these locations.

An important consideration is not, therefore, the value of a good, but its ability to pay transport charges, and this depends on its profitability considered apart from the costs of transport. Thus Lösch claimed that the transportability of goods did not depend on their value and that von Thünen's thesis that the 'lighter' or 'dearer' good was produced farther away from the market was false (Lösch 1954, 45; Hall 1966, 8), although it is not always clear from von Thünen's text whether value refers to gross return, price, or profit.

Changes in transport costs have resulted in changes in enterprise locations. Thus the growth of road transport has affected the location of fruit farming (Stamp 1943, 602–3) and of market gardening (Best and Gasson 1966, 53) and has brought a general extension of the margins of cultivation, together with a tendency at the national scale towards greater regional specialization (Chisholm 1962, 189–96 and see below p. 130). The modern agricultural map shows closer correlations between crop and physical factor distributions than ever before, and these new patterns are in a sense the product of changing transport technology. Such correlations may be expected to be more marked nearer the market than farther away owing to the tapered and stepped character of transport charges (fig. 7.1). Variation in the quality of transport available is sometimes neglected in enterprise location. Thus narrow roads in south-westernEngland are sometimes a limiting factor on the movement of bulk-milk lorries denying economies of scale to certain farms. In Breconshire in 1968 the enforcement of more stringent weight limits on eleven bridges made it impossible to move heavy loads to and from a number of farms until the bridges were brought up to standard, again raising costs or forcing a change of enterprise (*Farmers Weekly*, 22.11. 1968).

Some models of agricultural location in relation to the market

Von Thünen's model of agricultural location was based on the decline of economic rent or land rent with distance from market, and he applied marginal economics to the problem of cost substitution with increasing distance (von Thünen 1826; Grotewald 1959; Chisholm 1962; Hall 1966). Although it had a highly theoretical character and was presented in a deductive manner, nevertheless von Thünen's argument was derived largely from his experience of managing an agricultural estate near Mecklenburg. Essentially his model of agricultural location depicted a situation of partial equilibrium ignoring changes over time and assuming rational economic behaviour (Harvey 1966). It also assumed in its initial stages one high cost form of transport with equal accessibility in all directions from a centrally located and isolated market, uniform 'other' conditions, uniform scales of operation for a given enterprise or system of cultivation, and even competition for land between the different systems so that the more profitable systems per unit area were able to offer a higher price or economic rent for land than the less profitable systems and gain locations nearer the market. Von Thünen predicted a concentric series of agricultural zones or rings (fig. 7.2) around a central market, each with its own farming system and determined mainly by transport costs, weight and perishability of products, and the availability of manure. He developed a crop theory

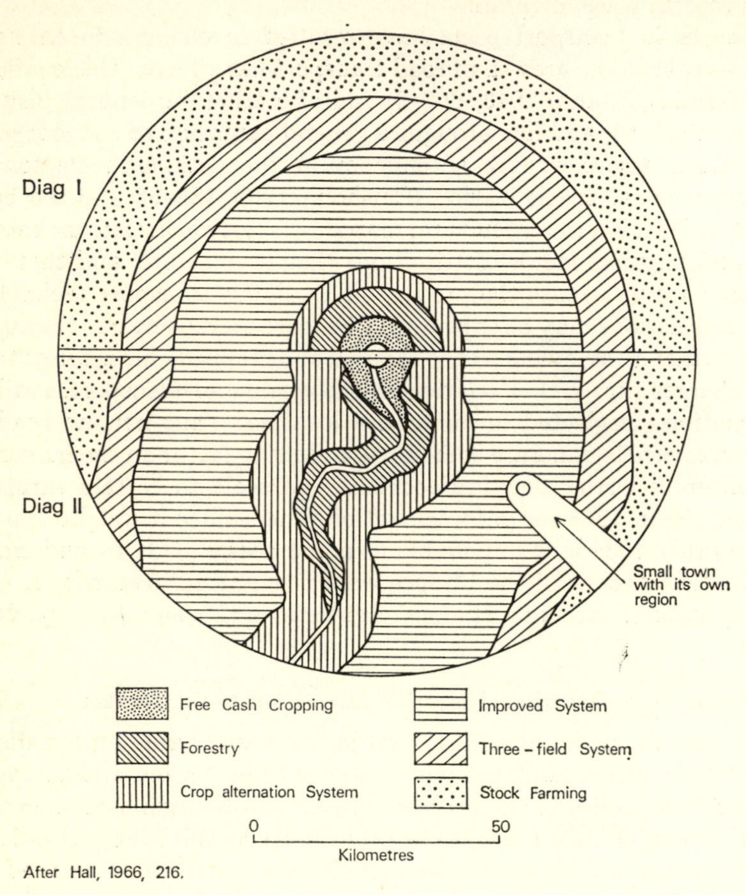

7.2 *Von Thünen's 'Rings' or concentric agricultural zones around an isolated town in a uniform plain.*

Diagram I General case.
Diagram II With navigable river and a small town with its own region.

and an intensity theory. The latter stated that intensity of production of a given crop would depend on the price the farmer received for his product less his transport costs and would, therefore, depend on distance from market. Intensity of cultivation therefore increased towards the market. His crop theory was concerned with the farmer's choice of crop with distance from market. This theory had something to do with intensity and was related to systems of production. It predicted, for example, that a more intensive system such as market gardening would be located nearer the market than a less intensive system such as livestock farming. Yet the relationship between crop pattern and intensity was by no means simple and involved considerable variations in yields and production costs. The impression that von Thünen's rings were solely determined by intensity was somewhat misleading even though his scheme consisted predominantly of systems of rising intensity towards a central market or town (Hall 1966, xx–xliv). Economic rent was assumed to be the product of transport costs or:

$$R \text{ (Economic Rent)} = E(p - kf) - A$$

Where A = outlay per unit area (not counting transport cost)
E = yield per unit area
p = market price per unit of yield
f = freight costs per unit of yield
k = distance from market (Lösch 1954, 38–42).

If we graph economic rent against distance (fig. 7.3) and assume regular relationships then we may conceive of two products with different gradients, that is different rates of decrease of economic rent with increasing distance from the market. The point where the gradients intersect indicates the margin of transference (see above) or distance from market of the boundary between the two products. Frequently those products least able to pay costs of transport were best grown under conditions of higher intensity, but there were several cost factors to be considered for each product. The analysis required to locate a whole system of production is extremely complicated, involving the costs of varying rotation systems, fallows, leys, labour inputs and yields. Thus, for example, the production of butter was excluded from locations near the market but not by its inability to pay rent. It had a low yield, high production costs and was more expensive to transport than grain, but it had a high value and its transport costs represented only a small proportion of total cost. Labour costs were the largest cost element and declined with distance from the market. Thus the decline in the production costs of butter was faster than the increase in its transport costs and grain was a better choice than butter near the market. Firewood and timber, compared with grain, had a high yield and low production costs. Given an expensive system of transport it appeared in the second ring, but often, as von Thünen admitted, towns were situated on navigable water and the big urban markets received their wood fuel or timber by canal or river from more distant locations. Livestock were a special case as they were the only farm produce capable of taking themselves to market, although in the case of fat cattle the loss of weight in droving

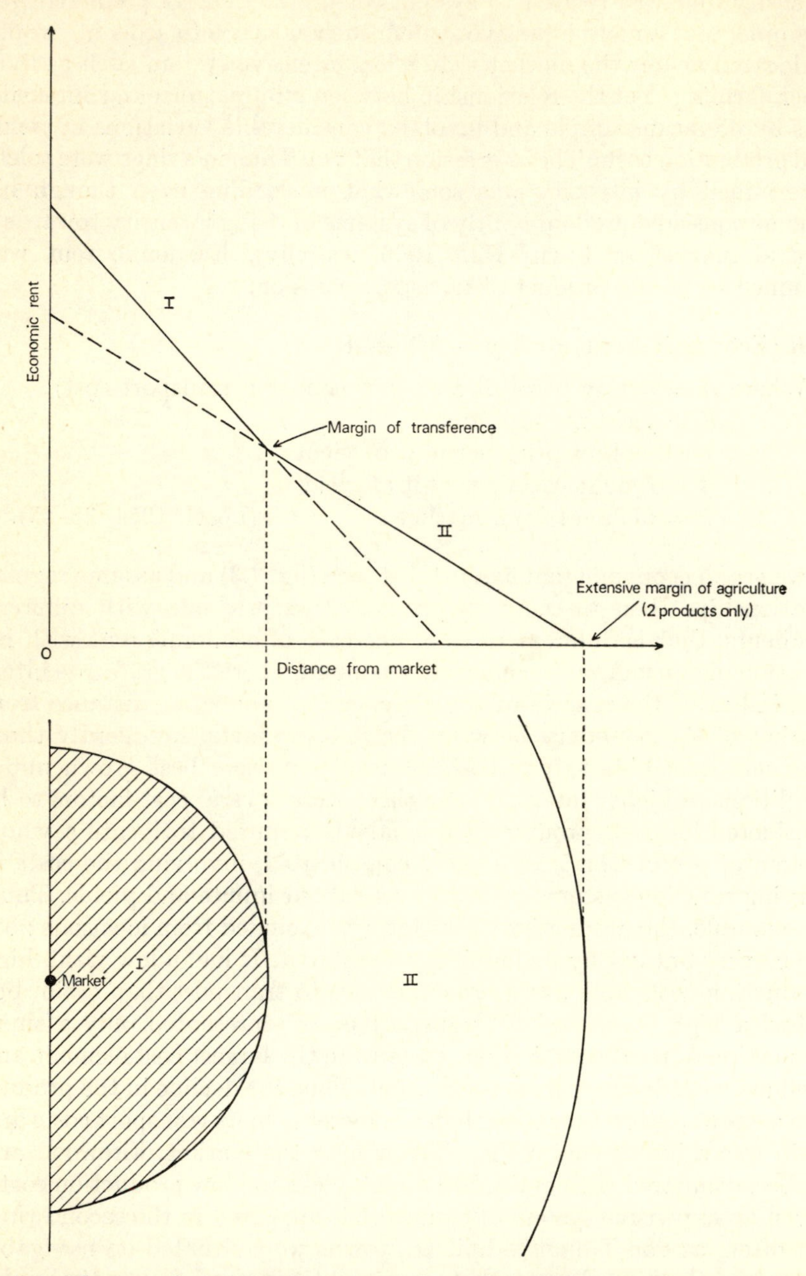

7.3 *Theoretical situation for two products (I and II) assuming constant but different freight rates.*

suggested the possibility of finishing, i.e. fattening on pastures, nearer the town.

Economic rent in von Thünen's model is not simply the difference in transport cost, but the difference in capacity to produce due to the difference in the costs of sending produce to market. Theoretically where other costs, such as transport, are lower, inputs can be raised proportionately further before corresponding returns begin to diminish. Other inputs therefore enter into economic rent, including, for example, labour and capital and were considered by von Thünen, more especially in his studies relating the 'Isolated State' to reality. We can hardly assume a uniform gradient for any one enterprise or production system in the graph of economic rent versus distance. We are more likely to suppose some stepping of the gradient with change of intensity and therefore change in the factors affecting economic rent. Probably the stepped line should be a curve rather than a straight line as with most enterprises proportionately higher inputs are normally required with each successive increase in unit of output. The tapering effect and stepped pattern of most transport charges would also tend to give the effect of of a stepped curve in the graph. Dunn has suggested the possibility of a curve and even the theoretical possibility of the appearance of a zone in more than one location or double margin of transference effect, although he doubted its 'empirical reality' (Dunn 1954, 43). Dunn expressed other misgivings:

1. That von Thünen's argument was at the industry level not the firm or farm level. It thus ignored the tendency of farms to increase production which they would do if other basic assumptions necessary to the model occurred, such as constant output, constant yield and constant price. Both competition and homogeneity of production were assumed, yet these were incompatible (Dunn 1954, 26–7).
2. It did not follow that the most intensive form of production would always be closest to market. It was possible to achieve a higher economic rent for a less intensive enterprise (Dunn 1954, 44).

This was, however, to some extent already implied in 'The Isolated State'. Dunn developed a much more elaborate analysis, examining the theoretical relations of costs, input and output under varying conditions, endeavouring to demonstrate that the rent function was non-linear. He also introduced, as did von Thünen, multiple markets in place of the single central market and variable transport rates, and discussed the effects of differences in physical resources and changes in population.

August Lösch argued that von Thünen's rings were special cases. Where two crops were raised the conditions under which crop 1 would yield a greater rent at the centre and a lower rent at the periphery than crop 2 ($R1 > R2$) were:

$$1 < \frac{E1 \,.\, p1 - A1}{E2 \,.\, p2 - A2} < \frac{E1}{E2}$$

that is, that the 'profit' (yield × price — 'outlay') of crop 1 is greater than that of crop 2, and the ratio of the 'profits' (dividing 1 by 2) is less than that of the yields, that is, that the proportionately higher yields of crop 1 involving higher costs of transport per unit area result in decreasing 'profit' with distance from the centre so that eventually the total profit (counting transport costs) of crop 2 becomes the greater (Lösch 1954, 40–2). Lösch observed that the more extensively cultivated crop was grown further out 'if only the smaller physical yield per hectare' was meant, 'independent of outlay'. He added: 'This condition is in fact necessary but not sufficient, since it is still possible that the extensive or intensive crop will be cultivated everywhere. But if "extensive" means smaller outlay per unit of area or of weight, this statement (that the more extensively cultivated crop comes from the outside) too is false' (Lösch 1954, 45–6). Lösch noted that with distance from a town other costs such as fuel, services, wages and actual rent (including the rent of labourers' houses) would be lower. This was due in part to 'the greater surplus of population' and to the lower density of population in the rural areas (Lösch 1954, 43). The availability of labour has been a most important factor in choice of crop and intensity of production, and the emigration of rural labour in the industrial countries has encouraged labour-saving innovations. In a situation where transport costs are of much less significance than labour costs we may even predict, were other things equal, the development of greater intensity of production at the more distant locations or inversion of the von Thünen rings assuming that machinery cannot normally increase the intensity of production. A similar effect may be produced by the development of short term leases at expanding city margins (Krueger 1959) or by other factors of urban encroachment (Sinclair 1967). We may note in this connection that an adjacent large city does not necessarily provide the chief market for a farmer. Horticultural produce, for example, is sold by some Vale of Evesham growers chiefly in the London market or in South Wales rather than in Birmingham. Elsewhere Best and Gasson (1966) have demonstrated a relation between urbanization, the decline of the small family farm and increased part-time farming. Although more part-time than full-time farmers operate at a very low level of intensity the authors suggest that in general the difference in intensity of land use between part-time and full-time farmers is quite small, part-time farmers merely preferring to operate simpler systems of production.

Other market location models were more empirical in character. Thus Jonasson in 1925 adapted the von Thünen model to the Europe of the 1920s (fig. 7.4) distinguishing a city zone with greenhouses and horticulture followed successively by market gardening, dairying, general farming, extensive grain farming, ranching and forest culture. He also found an almost identical distribution on the Edwards Plateau in

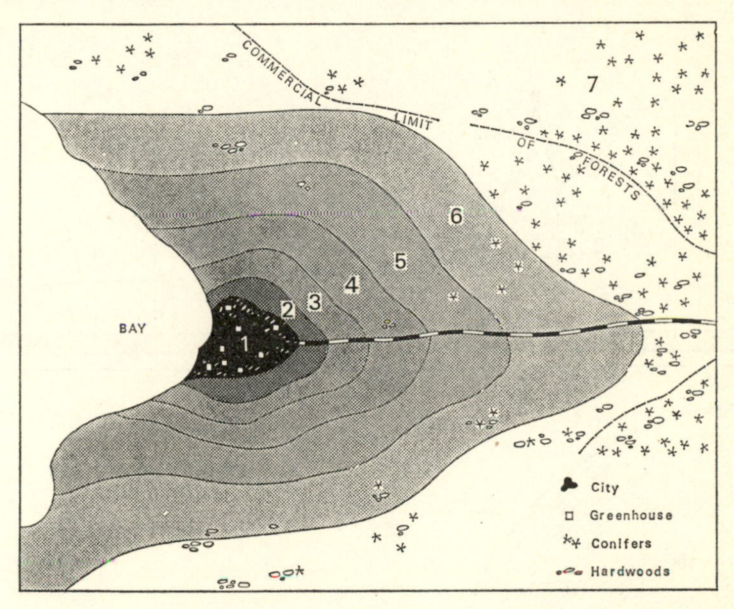

7.4 *Zones of production about a theoretical, isolated city in Europe according to O. Jonasson.*

Zone 1. The city itself and immediate environs. Greenhouses, floriculture.
Zone 2. Truck products, fruits, potatoes, and tobacco (and horses).
Zone 3. Dairy products, beef cattle, sheep for mutton, veal, forage crops, oats, flax for fibre.
Zone 4. General farming; grain, hay, livestock.
Zone 5. Bread-cereals and flax for oil.
Zone 6. Cattle (beef and range), horses (range), and sheep (range), salt, smoked, refrigerated and canned meats, bones, tallow, and hides.
Zone 7. The outermost peripheral area. Forests.

Texas (Jonasson 1925). In 1940 Cohen distinguished a series of zones, assuming transport costs to be the only factor. Beginning from the market she supposed the first would be the city followed by perishable fruits and vegetables including potatoes, then milk, wheat, butter and, finally, meat from cattle and sheep grazing on inferior grass (Cohen 1949, 36). Hoover in 1948 (fig. 7.5) constructed a model intended to indicate some of the diversity of land-use pattern to be expected in the

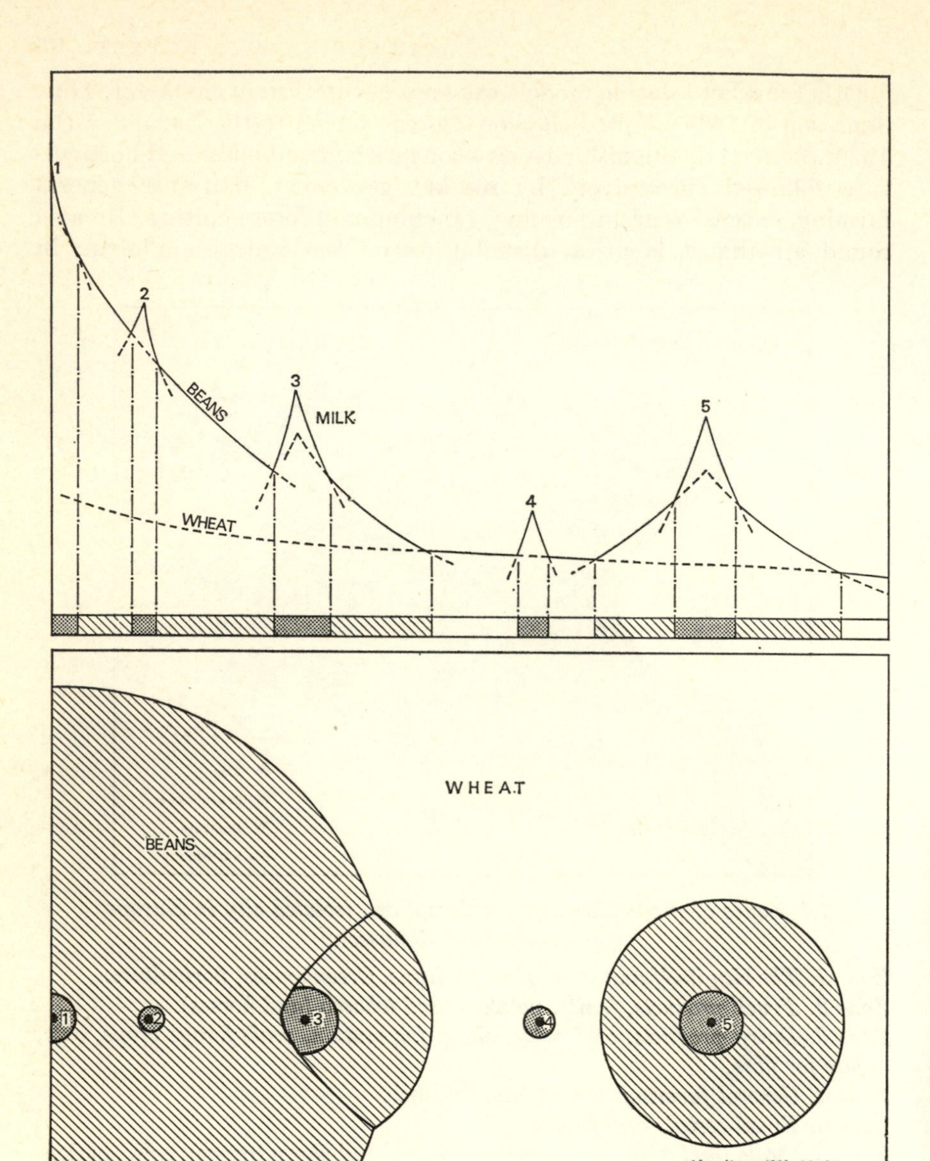

7.5 *Hoover's model of rent gradients and zones of land-use tributary to five market centres and involving three products.*

Rent gradients and zones of land-use tributary to five market centres, 1, 2, 3, 4, and 5. Three types of land-use are involved: milk, beans, and wheat production. It is assumed that a market for milk exists at each of the five market centres, that markets for beans exist at market centres 1, 3, and 5 only, and that a market for wheat exists only at market centre 1.

The upper part of the diagram shows the rent gradients for the three types of

[*Continued on facing page*

actual world. His model showed five market centres, each with a demand for milk, but with a wheat demand only at 1, and a beans demand at 1, 3, and 5. This 'complex pattern' for only three products 'under highly simplified assumptions' was meant to indicate why the patchwork observed in the actual world sometimes appeared utterly unsystematic (Hoover 1948, 95–7). Black constructed a simplifying model of the

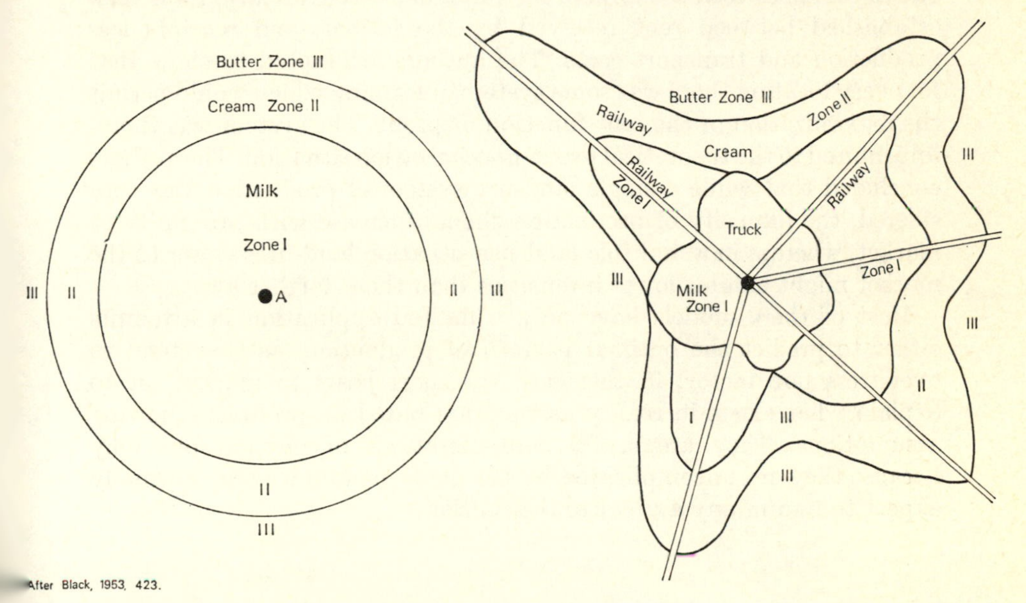

7.6 *Black's ideal and modified schemes of milk, cream and butter zones about a city.*

location of dairy production in relation to market distinguishing, successively from the market, milk, cream and butter zones, and showing the effect on these of the transport system (fig. 7.6; Black 1953, 423).

---

use, each rent gradient being drawn as a solid line over the interval in which the corresponding land-use is the highest rent-use.

The resulting progression of land-uses along the route running through the five market centres is shown by the shadings on the strip near the middle of the figure: stippled stretches of territory are devoted to milk production; shaded stretches to bean production; and white stretches to wheat production.

The lower part of the diagram is a map of the resulting pattern of land-use zones. Dots represent the five market centres. Stippled areas are devoted to milk production; shaded areas to bean production; and white areas to wheat production. The bean-supply areas of market centres 1 and 3 meet along the boundary that curves around market centre 3.

A more sophisticated study was made by Garrison and Marble who built an optimizer location model using an axiomatic system involving sets of crops, sets of markets, distance of markets from farms, yields, production costs, market prices and transport costs (Garrison and Marble 1957). This information or 'primitive notions' was used to state eleven axioms which in part acted as constraints on the system devised and in part described the form of the functions used. A relation was then established between rent received by the farmer, and receipts less production and transport costs. The authors attempted to show that for every location there was some system of farming which would permit the maximization of the rent function or profit. The system was therefore limited to the theoretical profit-maximization situation. The authors concluded that while a single land-use system of production was considered, the intensity of production should increase with proximity to market, whereas in a multiple land-use situation land-uses closer to the market might exhibit lower intensities than those farther away.

Most of these models have only a limited application in attempts either to predict the optimal pattern of production, as they tend to overstress one factor, in this case transport costs to market, or to explain what exists in reality as they are based on profit maximizing assumptions. They are useful demonstrations of method but only because they are uncomplicated by the many factors we may normally expect to find in any agricultural situation.

8 Government policy

The effects of government activity on the spatial organization of agriculture are discernible in almost every country in the world, and political rather than economic or social motives for policies are recognizable in many land-use patterns. Agricultural self-sufficiency is held by many governments to have strategic and prestige value, as well as acting as a means of satisfying the political demands of the rural vote. One of the more significant outcomes of the Kennedy Round (1967) of trade tariff negotiations, for example, is that trade in agricultural products will be the most difficult commodity sector to liberalize as national policies based on protectionist concepts are widespread. At the farm level the Government may limit the freedom of the farmer to select his cropping system. In some socialist states the precise rotation and combination of crops is stipulated for each farm. Under less authoritarian circumstances a more normal restriction may consist of quotas on the maximum acreages of particular crops a farmer is permitted to grow. In the same way the degree of control exercised by government over the marketing of produce also varies. In many countries, as in the United Kingdom, marketing boards exist as a means of regulating supply and demand and for the protection of numerous small producers. In many socialist states, however, most goods never reach the open market, the state purchasing products at the 'farm gate', and at predetermined prices. The whole structure of agricultural production may be dictated, as in the case of collectivization, but normally preferred systems are simply encouraged through the manipulation of financial support.

In most countries farmers benefit financially, at least in the long-term, from government activity in agriculture. It is in the interest of many governments to see that this is the case, and some have made strenuous efforts to stabilize prices and reduce uncertainty by protecting their industry from foreign competition. Nevertheless, government interference is resented by most farming communities. In Britain this resentment has been shown by the slow rate of acceptance by farmers of the National Agricultural Advisory Service. In developed economies most communities still cherish the concept of the family farm and its independent entrepreneurship and freedom of choice, which the Government by its actions is seen to be whittling away. While this is true to some extent, the loss of freedom of action resulting from large bank overdrafts and the development of horizontal and vertical integration is equally significant but less resented. Governments are seen as sponsors of change, collectors of taxes and removers of liberty; or, at their very best, as outsiders who lack understanding or sympathy for the demands of farmers.

Government policy depends largely on the level of agricultural self-sufficiency and the importance of agriculture within the national economy. On the basis of these two criteria most countries can be categorized into one of three groups. Firstly, the underdeveloped nations whose economies are largely dependent on the export of primary products. The major aims of their governments are to ensure self-sufficiency in food production where this is feasible, to stimulate cash-crop production, often with capital loaned from more prosperous nations, to obtain a reasonable price for their products in the world's markets and to absorb what surplus rural labour they can into industrial employment. Most governments have fought a losing battle. The terms of trade have moved against primary products since the early 1950s as world market prices for primary products have risen more slowly than for manufactured goods. Between 1953–5 and 1963–5, for example, an increase of 35% in the output of primary goods only represented an increase of 5% in the purchasing power of manufactured goods (Ojala 1967). Population increase has offset, or even overtaken, advances in food-crop production and the capacity of non-agricultural occupations to absorb surplus rural labour has hardly been sufficient. In an attempt to industrialize at maximum speed, many underdeveloped nations have ignored agriculture to an extent that they are no longer self-sufficient and now have to import food for their urban population. Others, such as Senegal, have had to import food because so large a part of their agriculture is concentrated on export crop production. Secondly, there are those industrially developed countries which protect their agricultural industries, not because cheap food cannot be imported but because they are unable to reduce the proportion of their food requirements they produce at home for political reasons. The major goals of these governments are,

(i) to keep agricultural incomes at a socially desirable level,
(ii) to obtain, in some cases, self-sufficiency in agricultural output,
(iii) to keep food prices to the urban, industrial consumer down, and
(iv) to increase the efficiency of agricultural production (O.E.C.D., 1965).

Policies (i) and (iv) are the most generally accepted, although they are often incompatible, at least in the short-term, as some governments adopt a policy of financially 'squeezing' their agricultural industry as a means of increasing its efficiency. Although the diversion of agricultural resources into other more profitable sectors of the economy is stimulated by such measures, governments are unwilling to endorse policies that would rationalize their industries completely, especially when the outcome could be politically disadvantageous. When translated into practical farming terms, the policy conflict over rationalization ultimately resolves itself into a choice between the sponsorship of large, mechanized, and vertically integrated units or the preservation, in some form, of the family farm.

Thirdly, there is a small group of highly developed countries, exemplified by the United States, which are embarrassed by food surpluses and excess production capacity. Technological progress has resulted in a recent rapid increase in output, at a time of inelastic internal demand for staple foods, particularly cereals. In order to remove this disequilibrium, raise product prices and increase farm incomes most governments have resorted to supply control measures, such as crop acreage quotas. These have only been successful where control has been introduced for all enterprises within a particular group, such as cereals, otherwise farmers have simply switched production capacity that is equally suited to the uncontrolled enterprises within the group and created the same problems afresh.

Most policy is formulated and decisions taken at the international, national, and regional scales.[1] Many national policies have regional implications, although these are not explicitly stated. For example, in Britain all hill farmers may claim extra subsidies under the Hill Farming Act (1946) irrespective of their regional location, but the subsidy obviously has a greater regional benefit for Welsh than for East Anglian agriculture. The possibilities of implementing a regional policy for British agriculture, and for Wales in particular, have been discussed (Bateman and Williams 1966). It was concluded that positive steps in this direction could be taken as only 9% of the subsidies currently awarded to Welsh farmers relate specifically to aspects of Welsh agriculture. Indeed, few nations have a regional policy for agriculture separate from more general regional planning considerations. Regional development plans, such as those drawn up by the Tennessee Valley Authority, do have spatial implications for agriculture, as farmers within a development area often receive special treatment. A good example of this is to be found in Southern Italy (*Mezzogiorno*) where a special fund (*Cassa*) was set up in 1950 to stimulate development in the

[1] Fielding (1964) has also described an example of more local political influence on the Los Angeles milkshed. He notes that political influences 'are associated with state (Californian) legislation that controls the production and pricing of milk, and with local land-zoning ordinances that facilitate the persistence of dairying near metropolitan centres' (p. 1).

South. Within the *Mezzogiorno* local authorities may obtain money from the *Cassa* for the improvement of their rural infrastructure and for large-scale land improvement schemes. At the same time individual farmers in districts designated as intervention areas (see fig. 8.1) may claim a grant of 60% towards the capital cost of farm improvements in excess of the ordinary grants available to all Italian farmers for farm improvement.

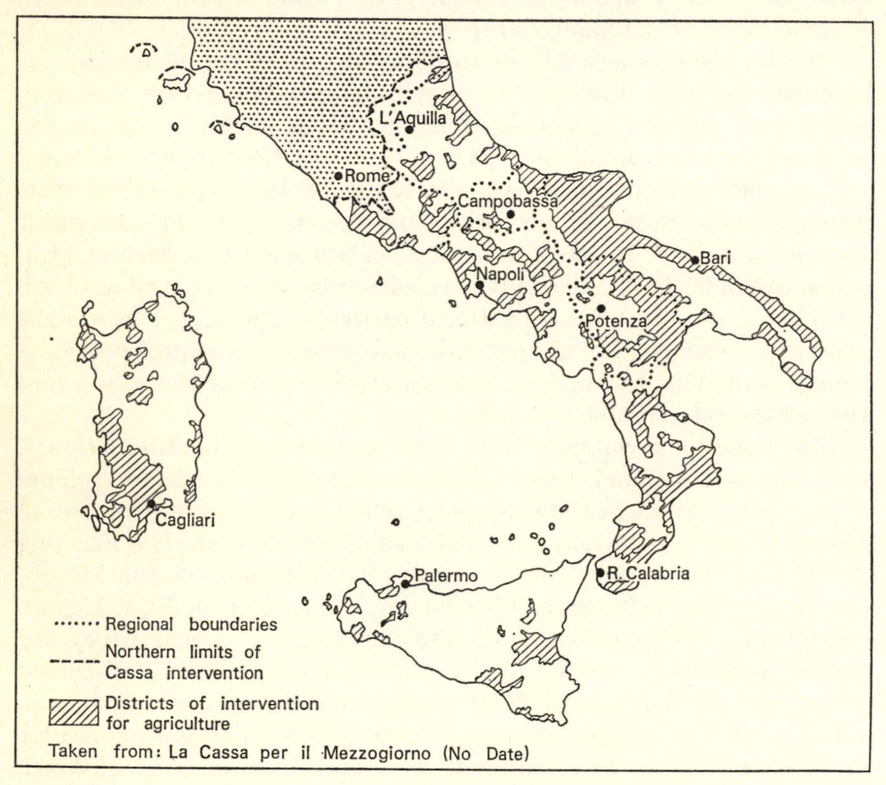

8.1 *Government intervention areas in the* Mezzogiorno.

International policies

The spatial imbalance of supply and demand of certain agricultural products creates problems beyond the resources of one nation to put right. Attempts to overcome such difficulties have been made either by interested countries agreeing on mutually compatible policies, such as commodity agreements, or by organizations such as F.A.O. or O.E.C.D., which promote agricultural development and provide channels for the flow of information, expertise and capital from the more to the less well-endowed nations.

The need for international commodity price agreements has long been realized, but because of the entrenched positions of individual nations, which fear the loss of established markets, policies have yet to be agreed for some products. There are at least two important reasons why such policies are necessary for the long-term economic development of nations dependent on the export of primary products, and both relate to the erratic nature of product prices in the world market. Firstly, demand from the developed world for many food products is inelastic and supply needs to be regulated if prices are to be stabilized. Secondly, it is necessary to establish prices high enough to provide underdeveloped nations with sufficient income to develop other sectors of their economies while not making it profitable for the developed world to grow or search for substitutes, such as sugar beet or synthetic rubber. Original attempts to control supply stemmed from unilateral action. The Stevenson Scheme (1922–8), for example, restricted rubber production in only the British Commonwealth countries and failed because producers outside the Commonwealth, such as the Dutch East Indies, continued to expand output. A similar failure can be recorded for coffee, and it has required adverse terms of trade to provide the stimulus for even some limited world-wide co-operation, as represented by the second Mexico City Agreement (1964) on coffee production quotas.

Preferential trade agreements are important influences on the distribution of certain commodities. Under the Ottawa Agreements Act (1932) preferential treatment was given to imports into the United Kingdom from Commonwealth countries. As the United Kingdom was, and still is, the world's largest importer of food this provided an important stimulus to the economies of Jamaica (sugar) and New Zealand (butter, lamb) in the aftermath of the Great Depression. The development of the E.E.C. is also resulting in major shifts in world trade, and some countries have even anticipated Britain's probable entry into it. Thus New Zealand is searching for new markets in south-east Asia, especially Japan, in order to reduce its dependence on Britain, and it is predicted that the E.E.C. agricultural policy will reduce the demand for hard wheat, whereas it will increase it for feed grains, from the United States (Fox 1967). This could accelerate the decline in U.S. wheat production, and wheat producers will have to examine the possibilities of substituting the production of feed grains for wheat. The Common Market agricultural policy will also lead to considerable internal adjustments in regional farming systems since the prices received by farmers in some areas will change considerably as the agreed common prices are adopted (Epp 1969; see fig. 8.2).

The locational effects on British agriculture of Common Market membership are at present uncertain. In general, the product prices received by the E.E.C. farmer are higher than those paid to producers in the United Kingdom. On the assumption that Britain

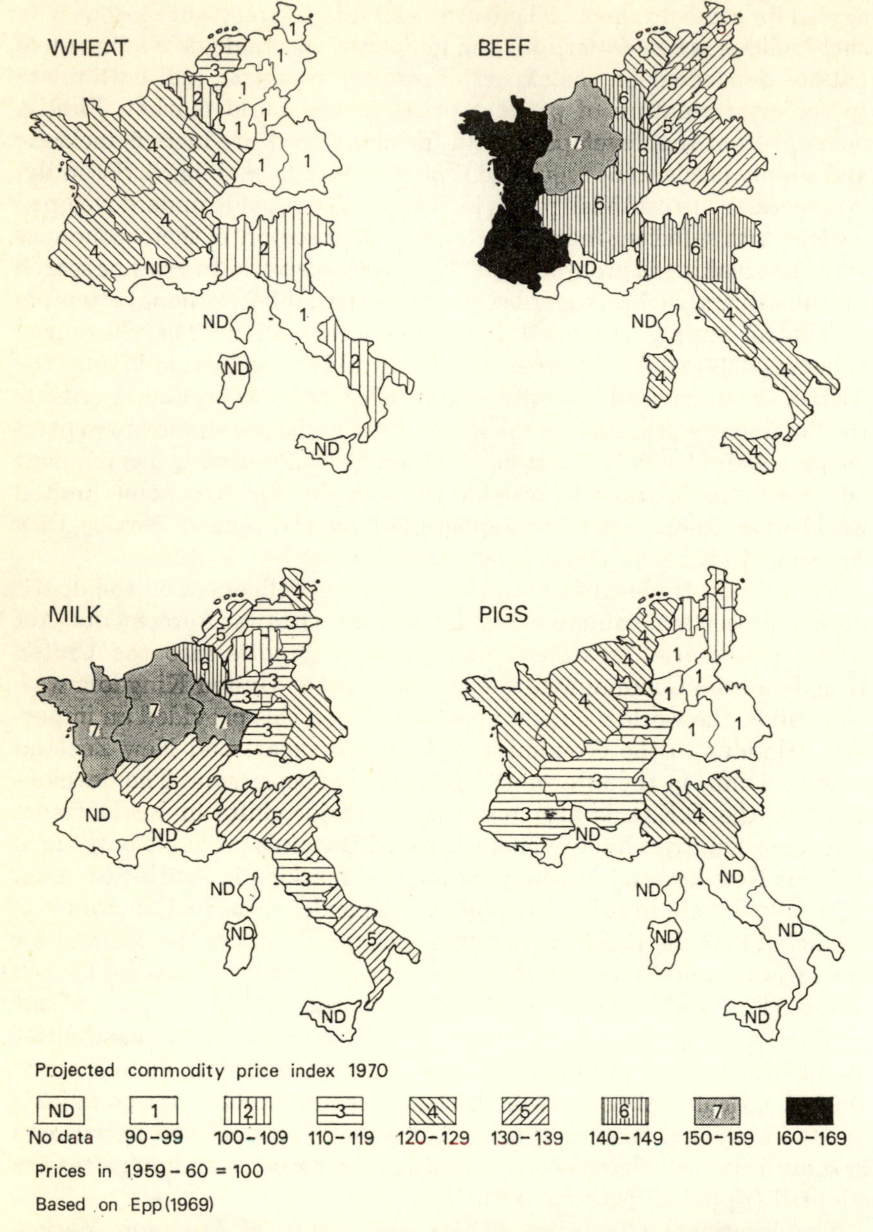

8.2 *Regional price adjustments (1959–61 to 1970) for selected commodities within the E.E.C. as agreed under the Common Agricultural Policy.*

will adopt the agricultural policy of the E.E.C.,[1] and while price adjustments will undoubtedly be made, the indications are that the United Kingdom grain farmer, particularly the wheat producer, will benefit most. Beef will command a higher price, but feed grain will cost more, while British potato producers and horticulturalists will probably suffer most. Egg and poultry producers in this country will face keener competition than at present, accelerating the rationalization of production into larger units, but the efficient producer could capture some E.E.C. markets (Donaldson and Donaldson 1969). The dairy farmer will have to become more dependent on grass because of higher feed-grain prices but until new forms of treating milk are generally adopted, should not lose any of the home liquid milk market owing to the current product's perishability. The major effect on the land-use pattern of Great Britain of these price changes will probably be to exaggerate the existing contrast between the arable east and the grassland west. Higher cereal prices will lead to an increase in cereal acreage, particularly in those parts of central England where mixed systems of production now predominate. Dairying is likely to decline further in eastern England as the economic advantages of continuous cereal production and the high cost of feed grains become apparent. In the west all livestock systems will become more dependent on grass and beef output will rise, particularly on those farms where capital has not been fully committed to milk production. Higher beef prices may also lead to an increase in output in eastern England where arable by-products can be used for feed purposes. Small farmers relying on pigs and poultry could well be forced out of business.

National policies

The intensely nationalistic approach of many governments towards their agricultural industries makes their internal policy decisions of paramount importance to the aspirations and expectations of farmers. Most significant is the extent to which the Government can reduce price uncertainty and allay fears as to the future status of agriculture within the nation's priorities. Four examples will demonstrate the degree of success normally achieved by democratic governments in influencing production trends.

[1] The major difference between the policies of the E.E.C. and the United Kingdom is that in the E.E.C. market prices are fixed at levels considered remunerative to E.E.C. farmers, usually well above world market prices, and these prices are protected by import tariffs. In the United Kingdom imports enter at world market prices and farmers are guaranteed minimum prices by the government. The cost of price support in the United Kingdom falls on the tax-payer, but in return he obtains cheaper food than in the E.E.C.

The Agricultural Marketing Acts of 1931 and 1933 permitted British producers to elect marketing boards, and these were set up for those products in which the nation was largely self-sufficient, such as potatoes, milk and eggs. One of the primary aims of the Milk Marketing Board was to reduce the transportation costs (10% of total costs, Chisholm 1957) of milk collection from the farm and its distribution from the dairy to the centres of demand. The centralization of milk transportation by the Board has led to a fundamental shift in the location of dairy farming and may in the long-term have increased the consumer price of milk.

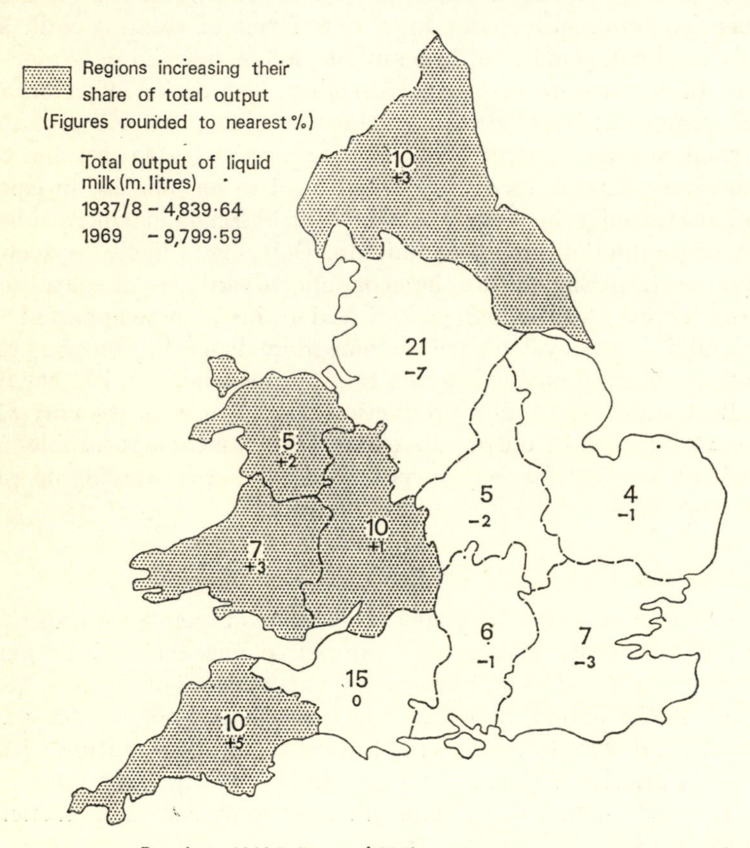

8.3 *Milk Marketing Board regions: percentage of total output of liquid milk in 1969 and percentage change 1937–8 to 1969.*

In 1933 the M.M.B. defined a number of regions for England and Wales within which producers paid a standard collection charge to the Board, irrespective of their location with regard to the nearest dairy. It has been estimated that up to 20% of the collection costs could be saved by ignoring the less accessible farms and the least concentrated areas of production (Chisholm 1957). Limited milk production in parts of eastern England, such as the East Riding, increases collection costs (route km/litre collected are greater) but not to the same extent as in upland western regions where producers have low outputs and road communications are poor. The differences in collection charges between the regions have also been minimized, and in 1970 the maximum difference in regional charge amounted to 1·7% of the producer milk price. This has largely removed the locational advantage of farmers close to centres of demand and has prompted the expansion of dairying in comparatively isolated parts of western England and Wales where the climate suits grass growth and arable farming is not particularly profitable (see fig. 8.3). A further policy has also stimulated this development. In the national liquid milk market supply exceeds demand and a much higher price is awarded to milk produced for the liquid market (4·12 p/litre or 18·72 p/gallon in 1970) than for the surplus used for butter, cheese and dried milk manufacture (1·88 p/litre or 8·75 p/gallon in 1970). The greater the national surplus the lower the overall price per litre paid to all producers irrespective of the fact that excess supply is greater in western regions (36% in North Wales) than in southern and south-eastern regions (less than 4%) (Simpson 1959). As a result the whole system has reduced the disadvantages of being far from either the dairy or the urban market and has encouraged dairying in areas such as upland Wales where production costs are high.

The collection and processing costs incurred by marketing boards in other countries have also not gone unnoticed. An investigation has been made, for example, into the possibilities of rationalizing the number and location of milk-processing plants in Munster and Co., Kilkenny (O'Dwyer 1968; 1970). An increase in size of collection centre reduces costs through a more effective use of labour and capital inputs, but as the size of plant increases so collection costs rise as the plant's hinterland has to be expanded in order to contain enough dairy farmers to provide the optimum throughput of milk (see fig. 8.4). In this case the optimum solution demanded a drastic reduction in the number of existing plants, from 140 to 23, as capital costs proved to be more significant than collection costs. Although the Irish Government controls all the marketing arrangements an immediate rationalization of the system to this extent would be financially impracticable and politically untenable as unemployment would result. On the other hand, the 393 egg-packing stations registered with the British Egg Marketing Board present an even more difficult situation. Not only do collection costs represent as much as 4½% of the retail price of eggs (Wilkinson 1969), but this cannot be easily reduced as the packing stations are privately owned. Consequently, all stations directly compete with each other, up to twenty stations operating in any one zone, and incur high collection costs as a result of route duplication (see table 8.1).

One of the most favoured means of encouraging production is the provision of subsidies, and their adjustment to stimulate or retard output. Experience demonstrates, however, that supply response to price change is not straightforward. There is always a time-lag between price change and response, and the Government can rarely hold all other relevant variables constant over this period. Response is always affected by the timing of the price-change announcement. A price increase in the February Price Review, for example, does not influence the acreage of winter wheat that year. Uncertainty also affects response. There was only a slow rate of acreage adjustment to grain price changes in Britain between 1925 and 1931, as this period was prior to government control of prices and one of price fluctuation (Hill 1965). On the other hand, response is not necessarily more predictable under more certain market conditions. Between 1956 and 1963 there was an exaggerated response in acreage to a slight increase in grain price, and this

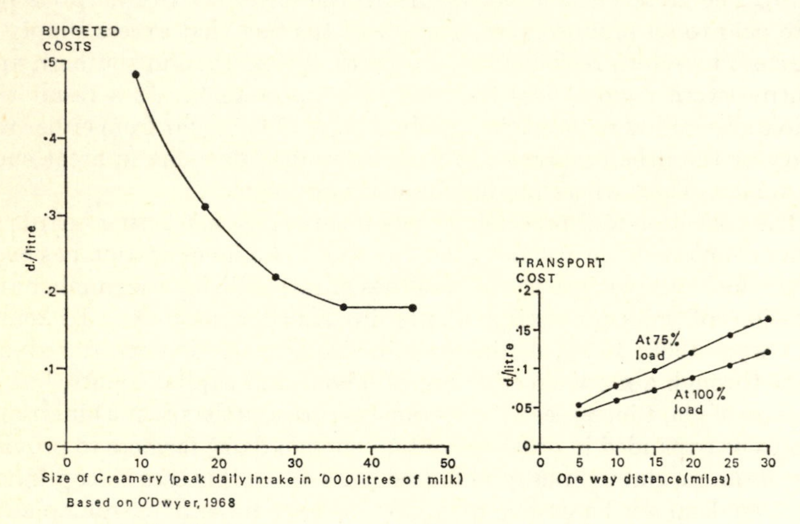

8.4 *Conflicting cost curves in the operation of creameries: an example from Eire.*

can only be explained by technological developments which led to a rapid increase in cereal yields during this period. A farmer can also select his system of production from a number of alternatives and, within certain limits, substitute one crop for another. Gardner (1957) has examined the response of the major British cereals to changes in their relative prices, on the assumption that it would be a simple matter for a farmer to substitute one for another. Seventy-two per cent of the

TABLE 8.1 *A selection of collection routes showing the mean values for each route of:*
(a) *Number of cases/pick-up*
(b) *Kilometres travelled/case collected*
(c) *Number of cases collected/man hour*

Route	*Number of pick-ups/ day*	*Total number of cases collected*	*Mean number of cases/ pick-up*	*Km*	*Km/ case*	*Cases/ man hour*
A	11	163	14·8	101	0·62	19·2
B	17	85	5·0	59	0·69	10·0
C	11	113·2	10·3	80	0·71	13·3
D	13	114·2	8·8	90	0·78	13·4
E	11	96·1	8·7	80	0·83	11·3
F	8	100·7	12·6	153	1·52	11·8
G	14	63·7	4·5	98	1·54	7·5
H	11	54·4	4·9	101	1·86	6·4
I	15	74·1	4·9	138	1·86	8·7
J	17	63·0	3·7	120	1·91	7·4
K	12	53·7	4·5	106	1·97	6·3
L	9	59·3	6·5	120	2·02	7·0
M	16	54·8	3·4	201	3·67	6·5
N	13	18·4	1·4	124	6·74	2·2
Weekly mean		2,839	9·2	2,673	0·94	

Based on: Wilkinson 1969, 334.

the variation in barley area was found to depend on the relative prices of wheat and oats but less than 40% of the variation in oats area on the relative prices of wheat and barley. This is because the function of the oat crop varies between east (cash crop) and west (feed crop) Britain, and 75% of the variation in oat area in west Britain could be related to the amount of stock feed imported the previous year. Similarly, it has been shown that the New Zealand wheat area is much more influenced by the relative prices of grass seed and fat lamb than the price of wheat itself (Fielding 1965). Policies that fail to account for crop or input substitution are unlikely to have much success in controlling output.

Since the mid-1950s a major agricultural problem in the United States has been over-production. By 1959 the surplus wheat holding alone had risen to 44,000 m^3 (1,200,000 bushels), more than one year's crop and twice the annual domestic consumption (Heady 1967). The distribution of surpluses abroad as 'food-aid' had failed to compensate for increasing yields, and supply restrictions, in the form of production quotas for certain major crops (wheat, rice, tobacco, cotton and peanuts) and the withdrawal of land resources, based on the 'voluntary land-retirement, compensatory payment approach', had to be introduced. The Soil Bank (1955–60) was an example of this. 11·6 m hectares were put into the Bank but only 4·9 m hectares less were harvested as some of the land put into the Bank had not been regularly cropped and non-participants increased their acreage in anticipation of higher prices resulting from reduced national output. Consequently, not nearly enough land was put into the Bank. Most of that taken out of production was also of low productivity as the compensatory payments failed to account in full for the higher incomes derived from better land. As a result, participation was highest in New Mexico and lowest in the Corn Belt (see fig. 8.5). The cost of compensation was, none the less, considerable, and an attempt to minimize future costs by optimizing the regional allocation of land withdrawal has been made (Whittlesey 1967). The results indicate that as much as 40–70% of all cropland in the southeast of the United States and nearly as much in the northern Great Plains should be abandoned. However, implementation would require the Government to enter the land market as it is inconceivable that the present voluntary approach would lead to such drastic land abandonment in the least-favoured regions. Needless to say, it would be politically untenable for the Government to endorse such a policy as it would result in large-scale unemployment in some rural areas.

Land reform, embracing changes in farm size shape, and fragmentation, as well as tenure, has attracted considerable attention and finance. The reasons for its sponsorship have been many, including economic advance, social justice and political gain. Radical changes in the demands made on rural areas resulting from population increase, advancing technology, and social and economic upheaval have meant that few governments have been able to ignore this issue. In the United States economists are advocating larger farm units as the only means of more effectively employing agricultural labour and capital resources, while in Latin America revolutionaries demand the break-up of large estates and the establishment of a landed peasantry as the basis for social justice.

Land reform has not always been successful. Governments have often failed to appreciate its dynamic nature and, in particular, the need to supply farmers with credit and advice after structural changes have been made. The reform inspired from below by Bolivian agricultural labourers in the 1950s, for example, failed to lead to an increase in food production, although a neo-feudal society was removed. Indeed, output fell by as much as 50% in the years following the reform, a situation reminiscent of the Soviet Union's collectivization programme of the 1920s. In Egypt and Japan, on the other hand, the immediate follow-up to land redistribution of advice and industrial assistance in the case of Japan, brought economic as well as social benefits.

The governments of the developed world have also taken action,

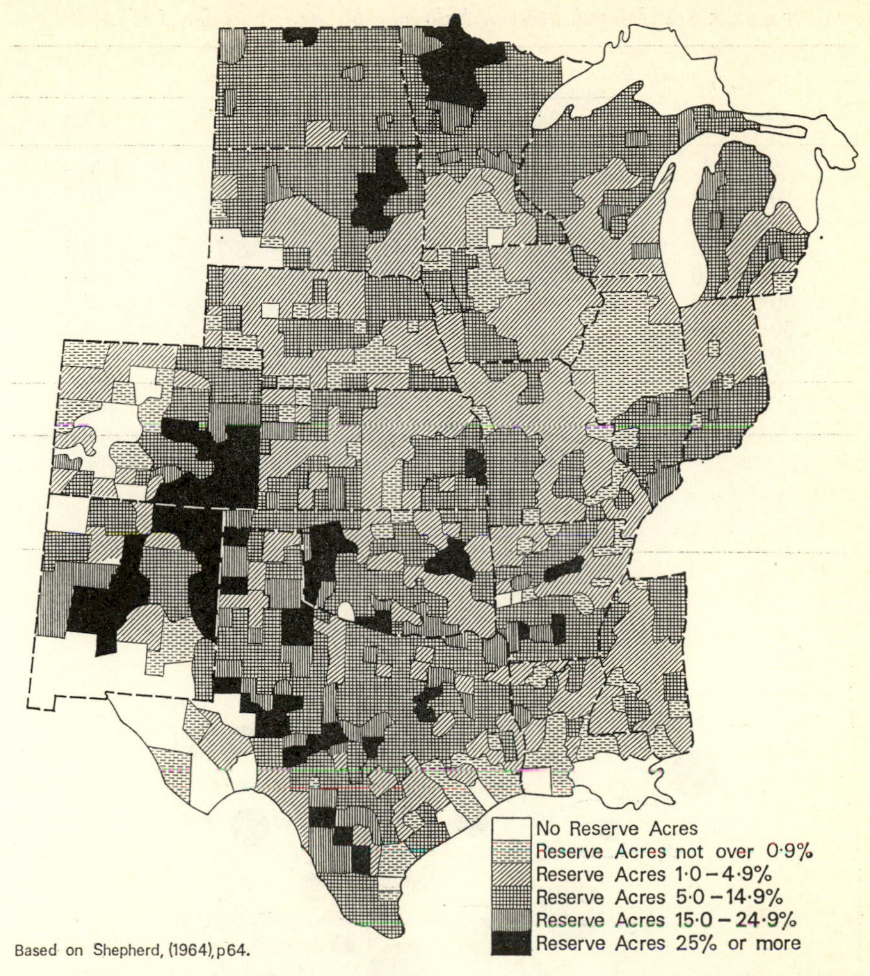

8.5 *Cropland in conservation reserve in the mid-west of the United States (1956–60 cumulated acreage on contracts as percent of total cropland in 1954).*

largely through efforts to increase the average size of farm and to improve security of tenure. In Sweden the County Agricultural Boards, on behalf of the Government, can purchase agricultural land when it comes up for sale. Only sales of land to the state or to a close relative are permitted without recourse to one of the Boards and these can refuse the transfer of land if the purchaser does not intend to farm it or the transfer fails to assist in the rationalization of the overall structure. By these means the Swedish Government has been able to increase the average size of holding by 30% between 1951 and 1966 (see table 8.2). The degree to which this policy has been successful is open to debate, however, for as fast as the structure has been improved technological

TABLE 8.2 *Distribution of holdings by size: Sweden 1951–66*

Size (ha)	Number of holdings (%) 1951	1956	1961	1966
2·1–5·0	34·0	32·7	28·6	25·4
5·1–10·0	31·8	31·0	32·2	29·5
10·1–15·0	14·0	14·5	14·2	14·0
15·1–20·0	7·2	7·7	8·7	9·5
20·1–30·0	6·3	6·9	7·9	9·9
30·1–50·0	4·0	4·4	5·1	7·1
50·1–100·0	1·9	2·0	2·3	3·4
100·1+	0·8	0·8	1·0	1·2
Total per cent	100·0	100·0	100·0	100·0

Taken from Whitby 1968, 285.

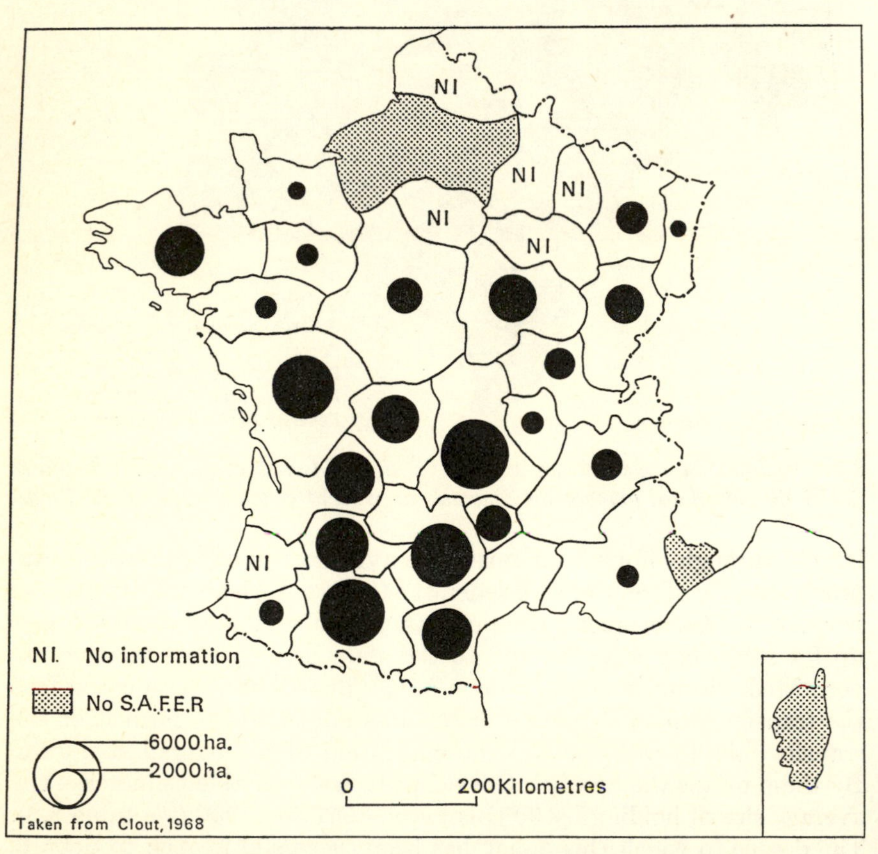

8.6 *Areas purchased by the regional SAFERs during 1966.*

change and economic conditions have increased the minimum size of the viable unit, and 50% of all land transactions since the Boards have been in existence have resulted from the initiative of individual farmers rather than that of the Boards. Furthermore, there already existed a trend towards fewer farms owing to poor agricultural employment prospects, an ageing farm population, and a slower rate of growth in farm incomes than in industrial earnings (Whitby 1968).

In France the Government has set up twenty-nine regional SAFERs (Sociétés d'Aménagement Foncier et d'Etablissment Rural) to acquire land for the creation of larger, more viable holdings. The extent of their activities is affected by the amount of land coming up for sale (limited in north-east France, more extensive in the Massif Central), the complexity of the tenancy arrangements (simple in southern France), and the price of land (high in Brittany and on the Mediterranean coast) (see fig. 8.6; Clout 1968). In no region, however, does the SAFER negotiate more than one-third of all land exchange and, as with the Boards in Sweden, the SAFERs are not likely to have more than a marginal effect on farm structure while it remains politically untenable for the Government to employ more extensive compulsory land-purchase powers.

9 Data and classification

All classification is subjective and to a considerable extent arbitrary. No 'natural' classification exists, i.e. in the sense that the data suggest incontrovertibly their own general classification (Harvey 1969, 331), although there may be 'natural' patterns or trends, even with a tendency for grouping or bunching of features here and there. Our data, selected with due regard to our imperfections of measurement and the effort and costs involved in the task, may be poor indices of these patterns. Classification enables us to order our information and therefore to obtain some glimpse of pattern, but it is only a glimpse, for the framework of classification, no matter how well we may devise it, obscures some of the pattern, or may even introduce some patterning of its own. The utility of a classification is related to the purpose for which it is designed, and its framework is therefore influenced by that purpose. Immense classificatory labours have all too frequently brought small result, and it has been claimed that maps based on classified data, such as maps of farm types, hide more than they reveal, so that geographers would be better employed in pressing for the regular publication of raw data (Chisholm 1964). A classification must be related to the problem to be solved. General classifications have frequently never been used, and if insufficiently flexible may even have inhibited research (Harvey 1969, 326–7). Classification faces not only problems of degree but also of kind, that is how to reconcile into one system different kinds of enterprise, involving differences in intensity of land-use or involving a chain of relationships so that a farm growing barley to feed cattle may be classified as a crop farm (land-use) or a livestock farm (product). It also faces the problem of placing the criteria in order of significance where farms may be defined by economic type, land tenure, enterprise combination, labour inputs, or other features, and where the hierarchical ordering of the criteria affects the result. Indeed, decisions about classification should precede data collection, since our collection depends on how we distinguish kinds of data, and, where a given kind is measurable, what cut-off points we establish to distinguish one degree or class from another. This latter problem is a particularly important one. Some geographers have argued in favour of abolishing classification by degree altogether and replacing it by ordination methods, using multivariate statistical techniques stressing gradient (Munton and Norris 1969) or trend surface analysis (Tarrant 1969).

Problems of data collection

The general problems of geographical data collection have been discussed by Harvey (1969, 350–69). The collection of original data by agricultural geographers is done at the farm or field level for both agricultural and land-use studies. The use of aerial photography as a source of information, despite the smallness of scale, may still in effect be regarded as a data source at the field level where interpretation is possible.

Unless frequent visits can be made to a farm, production data are only available for previous years, other than estimates based on experience. Time of visit is of enormous importance. Only in areas of extremely short growing season, say, three to four months, is it possible always to make clear distinctions between the activities of one year and the next. Often farmers may be preparing a field for next year's production before completing the current harvest in neighbouring fields. Some crops have longer growing periods than a year, e.g. manioc, others may stay in the ground but yield each year, e.g. bananas. Livestock enterprises have a wide range of periodicities and the time of sale of a beast may depend as much on future market prices as on the fattening plan, and may vary considerably between animals within a single system. Moreover, it is not always easy to distinguish the herd from the 'crop', i.e. livestock intended for disposal or sale, nor to be certain whether a herd of cattle is a dairy herd or a beef-producing herd, or both. Data collection may be carried out at a time of change when elements of two systems, past and future, are both present. The resultant analysis may in consequence fail to achieve a satisfactory study of location factors, and the data put alongside those of more stable systems may only cause confusion in interpretation.

It is possible to collect from just a few farms data of considerable scope and almost bewildering mass, yet to discover that the questions which they may be used to solve are extremely limited. Farmers are frequently reluctant to divulge the incomes they make or may expect to make from given enterprises. Sometimes they are reluctant to provide production figures or costings in case these may be used to calculate incomes. Often they do not even know in numerical terms, they just know that they are 'doing all right!' Yet since the main purpose of most farmers in choosing their enterprises is to provide for themselves and their dependants, even though they may not seek to maximize profits or may not know how to do so, knowledge of income is essential in attempts to explain locations. In some countries it is possible, having obtained production figures or estimates, to calculate approximate gross returns, providing market prices or dealers' prices are fairly constant. It is often possible to estimate or even obtain variable costs and thus obtain gross margins. Occasionally farmers can be found who are willing and able to provide the data for a reasonably accurate gross margin analysis. Such analysis is useful in comparing the performances of different enterprises on a farm and comparing the performances of different farms.

Geographers have frequently paid great attention to area data which are usually much easier to collect or to verify than data referring to input or output. We can obtain the area of each farm and allocate for a given moment or period of time the different proportions required by each enterprise. Problems arise where seasonal overlap occurs or in cases of mixed cropping. Area does not mention productivity nor plant population, for we would need to know the density. Moreover, we may also wish to know the use to which the crops, grass, or self-sown plants standing in the fields will be put, whether they will be sold or used for feeding, and if for feeding, to which livestock enterprises they will be allocated. A problem of this kind led Hartshorne and Dicken to conclude that grain production was more important than livestock close to the three largest cities of Europe, which appeared true on an areal basis and in farm-gate terms on the incorrect assumption that all the grain went to market and only grass hectares were counted as feed (Hartshorne and Dicken 1935; Chisholm 1964). In addition the amount of bought-in feeding stuffs presents a problem in using area as an index of business size. Moreover, the quality of land obviously varies greatly. A method commonly adopted is to use crude conversion ratios to produce an *adjusted area*, thus hill rough grazings may be divided by 6, upland rough grazings by 4, and permanent grass by 2, to relate to an arable base. Feed may also be converted into 'hectares' at an approximate rate of 1 tonne of concentrates equivalent to 0·32 hectares and 1 tonne of hay to 0·20 hectares (1 ton of concentrates to 0·8 acres; 1 ton of hay to 0·5 acres). Conversions for feed crops, where bought in, may be calculated from local yields. *Adjusted hectares* may produce a kind of index of the size of the farm firm and of its enterprises, but the index is extremely crude. Livestock enterprises may be similarly adjusted by means of *livestock units* or equivalents (formerly known as *cow equivalents*) (table 9.1). The livestock unit system, introduced as a method of data comparison and classification by Bennett-Jones in 1954, is one of standard equivalents weighted according to the estimated general feed requirements of selected classes of livestock. This estimation of feed requirements is based on Kellner's work on starch equivalents (Evans 1960) which provided a means not only of comparing one animal with another in relation to feeding stuffs but also of comparing the capacity of different foods to maintain an animal or increase its weight (tables 9.2 and 9.3).

TABLE 9.1 *Livestock units or dairy cow equivalents*

Grazing livestock	
Dairy cows	1·0
Bulls	0·8
Beef cows (excluding calf)	0·8
Other cattle over 2 years old	0·8
Semi-intensive beef (6–18 months old)	0·7
Other cattle 1–2 years old	0·6
Other cattle ½–1 year old	0·4
Calves up to 6 months	0·1
Ewes (including lambs under 6 months old)	0·2
Rams (over 6 months old)	0·2
Other sheep over 6 months	0·2
Sows (including litter to weaning)	0·5
Boars	0·4
Pigs fattened	0·1
Poultry over 6 months	0·02
Poultry under 6 months	0·005

Note that different livestock unit factors are recommended by different authorities. Each British university agricultural department recommends local variations for use in its own region.

TABLE 9.2 *Starch and protein equivalents of some feeding stuffs (per 100 kg)*

	Starch equivalent	*Protein equivalent*
Flaked maize	84	9
Maize	78	8
Barley	71	7
Decorticated groundnut cake	70	39
Decorticated cotton cake	68	35
Beans	66	20
Oats	60	8
Dried sugar-beet pulp	60	5
Molassed sugar-beet pulp	58	5
Barley, dried brewers' grains	57	19
Meadow hay (medium quality)	32	4
Barley straw (spring)	23	1
Potatoes	18	1
Barley, fresh brewers' grains	18	5
First-quality grass silage	12	2
Fresh marrowstem kale	9	1
Sugar-beet tops	8	1
Mangolds, yellow-fleshed globe	7	0·4
Rape	7	2

TABLE 9.3 *Food requirements of cattle*

1. Dairy cattle: maintenance (per day) according to live weight and breed

Breed	*Average live weight (kg)*	*(cwt)*	*Starch equivalent (kg)*	*Protein equivalent (kg)*
Dexter	300	(5·9)	2·0	0·18
Jersey	360	(7·1)	2·3	0·22
Guernsey	430	(8·5)	2·6	0·27
Ayrshire	480	(9·5)	2·8	0·30
Red Poll	510	(10·0)	2·9	0·32
Dairy Shorthorn	530	(10·5)	3·1	0·33
British Friesian	580	(11·5)	3·3	0·35
South Devon	660	(13·0)	3·5	0·39

2. Production (per litre of milk) (=0·22 gallons)

Breed	*% fat in milk*	*S.E. (kg)*	*P.E. (kg)*
Jersey	5·1	0·3	0·07
Guernsey	4·6		
South Devon	4·3	0·28	0·06
Dexter	4·2		
Ayrshire	3·8	0·25	0·05
Red Poll	3·6		
Dairy Shorthorn	3·6		
British Friesian	3·5		

3. Beef cattle: nutrient requirements per day for liveweight gain

Live weight (kg) (1 cwt = 50·8 kg)		*MaintenanceS.E. (kg/day) (1 lb = 0·45 kg)*	*S.E. required in addition to maintenance (kg/kg liveweight gain)*
50–150		No standards	1·1–1·5
150–250			1·5–1·8
250–325		2·0	1·8–2·0
325–375		2·3	2·0–2·3
375–425	Store	2·5	2·3
	Fresh		2·5
450–500	Store	3·0	2·5
	Fresh		2·8
	Half fat		3·0
	Fat		4·0

Protein requirements are much more difficult to estimate. Generally 0·6 kg/day are required for the maintenance of a 300-kg bullock with a liveweight gain of 1 kg per day, while 0·7 kg are required for a 450-kg bullock with a liveweight gain of 1 kg per day. If, as in store feeding, the liveweight gain required is only ½ kg per day, then less protein is required, say, a ration of 0·5 kg per day for a 450-kg bullock.

TABLE 9.3 (*continued*)

4. Summer grazing involves energy expenditure varying according to the quality of the pasture

Type of pasture	*S.E. used in the effort of grazing (kg/day)*
Good fattening pasture	0·5
Average pasture	1·0
Poor pasture	1·5

Note that many pastures deteriorate, accumulating mature coarse growth in the herbage during the late summer and autumn.

Based on Evans 1960, 45–69.

In table 9.4 some simple calculations are provided to show the kind of use to which starch equivalents can be put, in this case the comparison of barley and pasture grass as feeding stuffs in terms of areas required. These figures, for simplicity in presentation, ignore other requirements such as protein.

A similar energy measure is provided by the use of joules. By relating production to human consumption a rough indication of the capacity of a given system to support human life can be obtained. Stamp's Standard Nutrition Unit of 4·2 million kJ per year (or 1 million kcal.) (equivalent to a daily consumption of 2460 kcal. plus an allowance for waste in preparation, cooking, storage and transport) allows us to make a very crude evaluation (Stamp 1958 and 1960, 115–19). It is, however, essentially a measure as much of management as of choice of enterprise and of land quality.

Calculations of this kind provide crude standard indices or guides. They can never replace actual production figures obtained from individual farms. Figures of output or input whether in energy or value terms are useful not only as measures of the firm or system but because they provide a single measure for both crops and livestock. We may use financial conversion ratios or labour conversion ratios to convert our areal or land-use measures into standard indices of input or output. In countries such as Britain elaborate standard input and output performance figures and conversion factors are calculated every year, often varying regionally (e.g. Nix 1969). These standards have been used by agricultural economists and farmers as guides to production efficiency. They may also be indicators of expected farm business inputs or outputs making certain assumptions about farm size, business efficiency and land quality, and have been used for classification purposes by both economists and geographers (table 9.5); Jackson *et al.* (1963) have used output data in the form of gross margins to classify

TABLE 9.4 *Comparison of barley and grass feeding areas required by Jersey cattle*

Theoretical requirements for a Jersey cow weighing 360 kg (7·1 cwt) to produce 4100 litres (900 gallons) of milk per year (based on starch requirements only and making no allowance for growth, pregnancy, or calf)

1. Feeding stuff requirements (basis as winter milk production):

Maintenance (per day) 2·3 kg	=	840 kg (per year)
Production (per litre) 0·3 kg	=	1230 kg (per 4100 litres)
		2070 kg (starch equivalent or S.E.)

Barley equivalence (71 kg S.E. per 100 kg feeding stuff):

$$\frac{100}{71} \times \frac{2070}{1} = 2915 \text{ kg of barley required}$$

Assume barley yield of 3760 kg/ha (30 cwt/a.), then area of barley required to feed Jersey cow as above is 0·78 ha (2 acres approx.)

2. Grass pasture requirements (basis as summer milk production):

Maintenance (per day) 2·3 kg	=	840 kg (per year)
Production (per litre) 0·3 kg	=	1230 kg (per 4100 litres)
Required for grazing effort (on good fattening pasture per day) 0·45 kg	=	165 kg (per year)
		2235 kg (S.E.)

Utilized Starch Equivalent (U.S.E.)/hectare range: 1000–4000 kg (900–3600 lb/a.) (based on Evans, 1968, 104). Assume average of 2500 kg U.S.E. (2200 lb/a.) available, then area of grass required to feed Jersey cow as above is 0·89 ha (2·2 acres).

Note: These comparisons are purely theoretical and are intended solely to serve as indicators of the effectiveness of different forms of land use. They do not allow for seasonal and other variations, nor do they allow for the normal practice of providing cattle both with grazing and feeding stuffs.

different farm systems in eastern England. Generally farm input data, usually man-hours or man-days (table 9.6) are preferred because inputs are held to reflect the intentions of managements better than outputs (Jones 1965) and labour figures tend to change more slowly than money values (Ashton and Cracknell 1961). This method has, however, been criticized as overweighting labour-intensive enterprises and failing to allow for seasonal variations in the opportunity cost of labour (Belshaw and Jackson 1966). A combination of area proportions and livestock

TABLE 9.5 *Standard gross outputs* (£)

Crops (*per acre*)		*Livestock* (*per head*)	
Wheat	40	Dairy cows	125
Barley	35	Barley beef	70
Oats	35	Beef (6–18 months)	45
Potatoes	130	Other cattle over 2 years	25
Sugar beet	90	Other cattle 1–2 years	25
Vining peas	80	Other cattle ½–1 year	12
Threshed beans and peas	30	Calves	25
Herbage seeds	40	Ewes	8
Hops	400	Other sheep over 6 months	5
Hay for sale	15	Sows	75
Keep let	10	Other pigs over 2 months	28
		Laying birds	2
		Pullets reared	0·7
		Broilers	0·2
		Turkeys	1·2

Taken from Nix 1969, 122.

TABLE 9.6 *Standard man-days* (*1 S.M.D. = 8 labour hours*)

Crops (*per ha*)		*Livestock* (*per head*)	
Wheat	3·7	Dairy cows (parlour)	7·5
Barley	3·7	Dairy cows (cowshed)	10
Oats	3·7	Bulls	4
Potatoes	37	Beef cows	2·5
Sugar beet	25	Barley beef	3
Vining peas (mobile viner)	5	Beef (6–18 months)	3
Threshed beans and peas	5	Other cattle over 2 years	2·5
Herbage seeds	2·5	Other cattle 1–2 years	2
Hops (machine-picked)	136	Other cattle ½–1 year	1
Hops (hand-picked)	309	Calves	1
Feed roots (cut)	30	Ewes	0·75
Feed roots (folded)	7·5	Rams	0·75
Kale (cut)	15	Other sheep over 6 months	0·5
Kale (grazed)	2·5	Sows	4
Hay/silage	3·7	Boars	2
Keep let	1·2	Other pigs over 2 months	0·75
Bare fallow	2·5	Laying birds, intensive	0·1
		Laying birds, free range	0·25
		Pullets reared	0·05
		Broilers	0·01
		Turkeys	0·1

Taken from Nix 1969, 122.

numbers has been used but is complicated by the employment of two measures. On the whole preference probably goes to financial data as a measure of farm-firm type with area × financial conversion factor as a poor substitute. In most countries such data are not available or are only available for large unit areas and not for individual farms. The United States probably has the best provision of both farm and large unit area data for the agricultural geographer of any country in the world, and it is there that some of the most advanced work has been done on farm classification, enterprise location and regional agricultural analysis (see, for example, Olmstead and Manley 1965). For comparisons of production and consumption in countries with peasant agriculture, grain equivalents, which relate production to market value, have been used (Buck 1937; Clark and Haswell 1964, 48–68). One of these, the wheat equivalent, is based on a median of regional wheat relative price weights (Klayman 1960) using conversion coefficients for each crop and for livestock products. It is a device which may provide some crude general comparisons, but which necessarily obscures enormous variations in the data on which it is based.

Sampling

Few geographers or economists can visit more than a hundred farms in a year's work, collecting a large number of variables from each farm, allowing a day per farm for basic study, and another day for a second visit later in the year to sample changed seasonal conditions and to check some of the initial data and results. Agriculture, especially in Europe and North America, is currently subject to rapid change, and in most studies it would be difficult or impossible to incorporate the results of two or more season's work into a single programme. In consequence, unless large teams of workers are used, sampling is required to cover areas sufficiently large for the discovery of general trends or patterns.

The 'representative farm' or case study became popular in the 1920s for studies of types of farming and was in effect a crude form of sampling, involving an appreciation of farm practice in a given area and a highly subjective judgement of the fairness of the sample. The method is useful for the demonstration of relationships on actual farms, but is of limited or no value for generalization. It is still used in current school texts and has been used to illustrate general regional or world descriptions of agriculture (e.g. Dumont 1957). Combined with a general survey the study of a 'specimen farm' may provide a useful insight into the operation of local farm systems (Henshall 1967).

A sample survey may sacrifice the detail to be gained from case studies in order to achieve statistical adequacy in the study of a region or area. In consequence Blaut argued that there was a need to combine microgeographic research procedures with formal sampling design in order to achieve a high intensity of research over a wide area (Blaut 1959). He distinguished:

THE SIMPLE RANDOM SAMPLE

This may be taken from a list of farms by drawing lots or using a table of random numbers, and all farms, irrespective of size, have an equal chance of being selected. A regular sample, i.e. a sample at regular intervals, can be taken, providing one can be sure that there is no pattern or regularity in the list. The random sample may be taken from an area by use of a grid. Grid intersections can be chosen by random numbers and the farms identified where each intersection occurs. Such identification depends on field work to establish ownership. The method may have the disadvantage of bias in selection towards larger farms, although it has been argued that this could be a useful principle in farm sampling in order to distinguish 'the relative importance of farms of various acreages' (Belshaw and Jackson 1966). If importance is directly related to area it is a useful principle, although Belshaw and Jackson have admitted that where land quality differed widely it would 'probably be necessary to give less weight to poor land'. An even spacing sample has been used in a survey of farming in the Isle of Man (Birch 1954 and 1960) in order to obtain an adequate spatial distribution of farms of all sizes while giving a greater proportional representation to the larger holdings. Where peasant farming is involved sampling procedures are often essential for the survey of large areas, but may be difficult or impossible to organize on a satisfactory basis where no list of farmers and no map showing farms are available. Sampling frequently has to be on a village basis and farmers chosen in each village after consultation. Procedures frequently involve finding a 'typical' village after discussion with government agricultural officers, and then finding farmers willing to co-operate. Sampling is therefore seldom strictly random and considerable bias is possible. Upton's study in south-western Nigeria provides an excellent account of the difficulties (Upton 1967).

THE STRATIFIED RANDOM SAMPLE

Sometimes the objectives of study are more concerned with understanding types of farming than with obtaining an even areal picture. In such a case it is useful to stratify the sample and to select farms randomly in each of the strata. Such a technique requires advanced information and can introduce a serious subjective bias where advanced information is inadequate.

THE CLUSTER OR BLOCK SAMPLE

This involves the random sampling of areas rather than of farms. The area to be studied is divided into equal-sized units, usually grid squares, and these are sampled by randomizing procedures. All farms in each square are studied. Locally, detailed variation may be discovered, but large areas may be left untouched, making mapping of the results impossible. Variation in response rates within each square may make comparison difficult.

Use of data classified by areas and not by farms

Official data, usually collected in agricultural censuses, intended to assess production and assist in planning agricultural policy and development programmes, are rarely available to research workers as individual farm returns. It has been argued that they would not greatly improve our ability to explain farming patterns if they were, as they simply tabulate information and do not inquire into the motivation of the farmer (Munton and Norris 1969). However, there are some advantages in countries such as Sweden, where both the boundaries of holdings and the size of individual incomes can be ascertained, or Hungary, where a much greater wealth of economic data is available than in Britain (Coppock 1968). Published information usually refers to administrative or other areas used for data collection. Such information has the advantage of covering large areas, of offering great savings in time and labour, and sometimes of offering data for the same area at regular intervals of time so that changes may be studied. In Britain the smallest areal unit of statistical information easily available is the parish, but this has been found to be too small for the construction of an agricultural atlas of the whole of England and Wales (Coppock 1964*a*). Unfortunately most data-collecting units provide an awkward frame of reference due to variation in size and the lack of coincidence between their boundaries and the boundaries or patterns of the feature to be studied. The 12,000 parishes of England and Wales, for example, range in size from 9 to 25,609 hectares, have no simple areal distribution by their different sizes, and have boundaries which do not coincide with farm boundaries so that the parish data do not normally refer precisely to the parish area. In this respect the parishes of England and Wales are less satisfactory than an American county (Coppock 1960). In some parishes more than half the land belongs to farms recorded as belonging to other parishes. Some farms are highly fragmented with a wide scatter of fields. Farmers may buy or rent land at some distance from their main farm in order to acquire extra pasture or more work for their machines. Holding amalgamations have taken place and two or more widely separated holdings may be farmed as one unit. Occupiers of multiple holding units have been encouraged to make a single return provided that the holdings are farmed as one unit and in the same county (Coppock 1965*b*). Transfers of land involving changes in size of holdings make it impossible to use parish data for the assessment of agricultural change unless information on land sales or tenancy changes is also available. The comparison of agricultural distributions with other distributions, such as soils and relief, is particularly suspect because of the wide variation of physical conditions in most parishes. The use of larger areal units for broader generalization or smaller units such as farms, or even fields, can assist understanding, but correlation is only to be understood for the area concerned and the factors will have different proportionate importance at the different scales. For this reason Chisholm has rightly attacked the fallacy of composition, i.e. the notion that work involving the assessment of location factors can begin with small regions which can then be added together into larger units for generalization over larger areas (Chisholm 1964).

Where, as is frequently the case, there is wide local variation in the proportionate importance of enterprises on farms, the lumping together of data in frameworks which ignore this local variety not only brings errors of location to any interpretation but also major errors in classification or in the understanding of the relationship of the enterprises involved. To record the proportions by area, labour inputs or value of the enterprises in a given district tells us little or nothing of the agricultural systems which occur there. While enterprises may be combined on a farm for some joint advantage, enterprises which appear together in a single district may have no interrelationship and may even be the result of combining data for two or more specialist farms, each with a single enterprise. District associations may, however, offer some notion of local enterprise predominance. The use of equal-sized squares for data collection and mapping facilitates areal comparison, although clearly there are considerable problems of boundary overlap and data lumping. Such squares have been used for the construction of 'Types of Farming' maps of England and Wales, using a 10 km grid and sampling one-sixth of holdings (Ministry of Agriculture 1969; fig. 9.1).

Classification of types of farm

The most general classifications of farm types are those based on a general appreciation of the purpose of the system or the enterprises involved. Thus frequently distinctions are made between cash-crop farms and subsistence farms, between peasant farms and plantations, between shifting and sedentary agriculture. Farms may be classified according to the extent to which they appear to resemble the model type, although the type itself may be difficult to define. Often it is a vague idea for which new definitions are constantly being supplied. Sometimes these reflect changing economic conditions. For example, Gregor has sought a better definition for the modern plantation, claiming that many of the supposed characteristics assumed to belong to it are now becoming associated with other types of farm (Gregor 1965). Others have sought

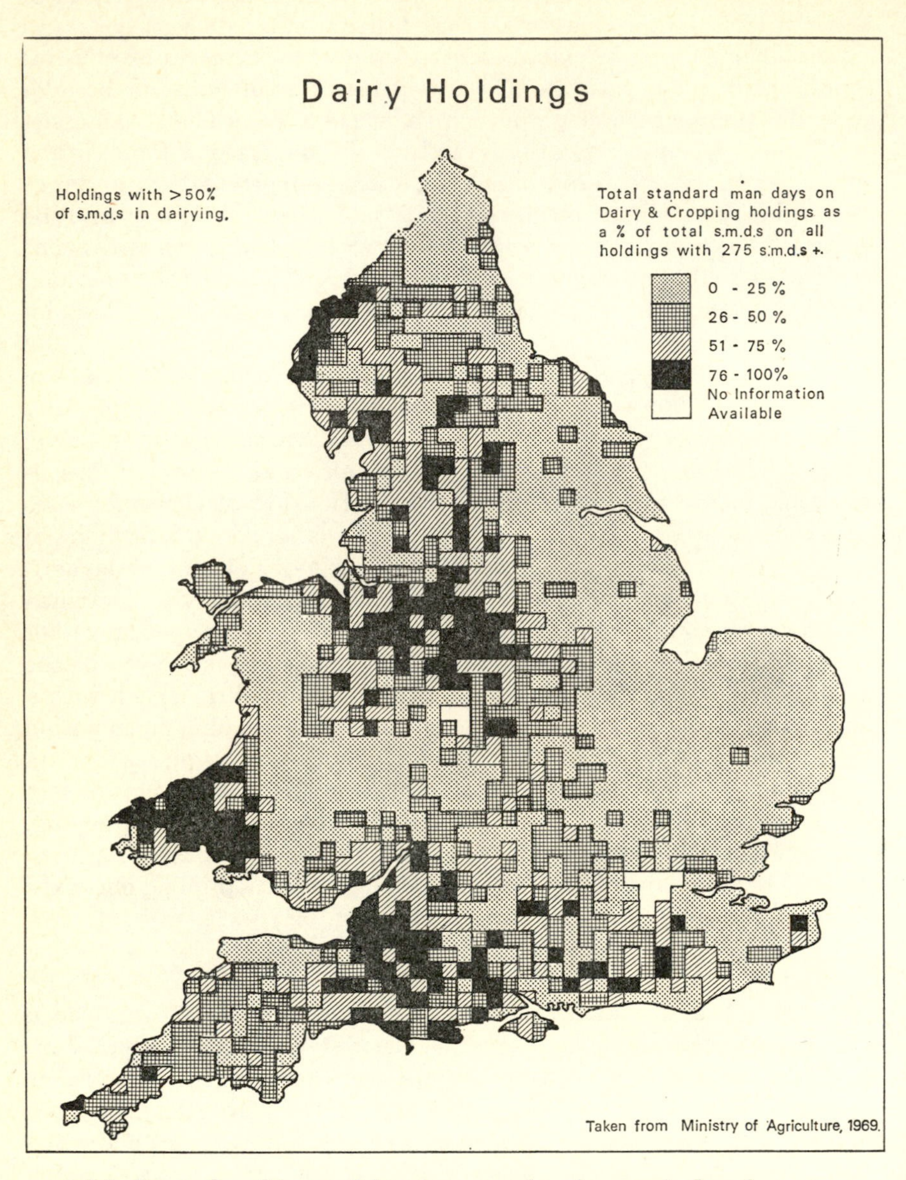

9.1 *Examples of types of farming maps based on agricultural census data for 1965.*

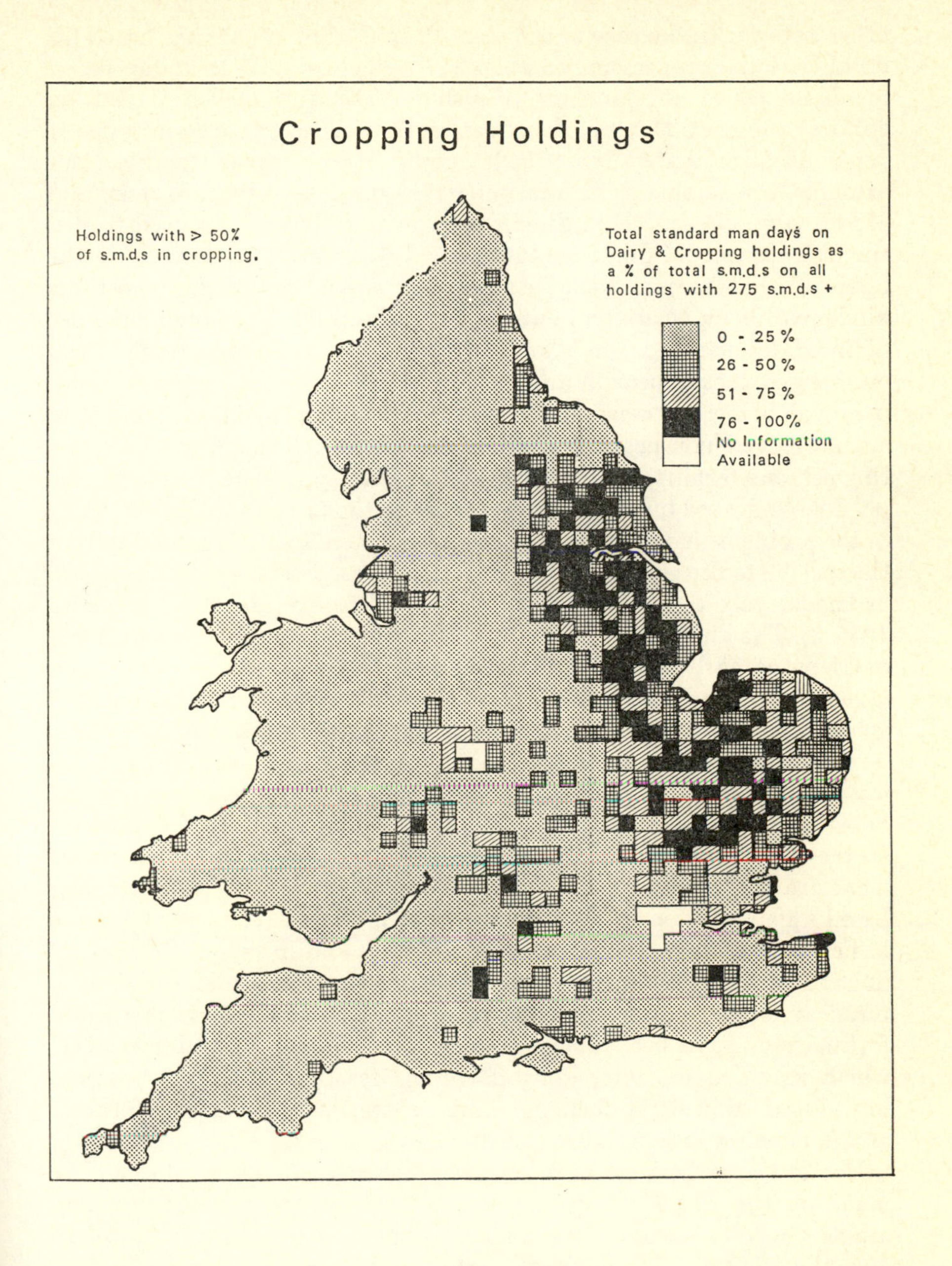
Cropping Holdings
Holdings with > 50% of s.m.d.s in cropping.
Total standard man days on Dairy & Cropping holdings as a % of total s.m.d.s on all holdings with 275 s.m.d.s +
0 - 25 %
26 - 50 %
51 - 75 %
76 - 100%
No Information Available

other criteria. In the case of the plantation Gerling (1954) has based his classification on processing complexity, while Steward (1960) has based his on stages of development (Henshall 1967) and Jackson (1969) on cultural affinities. Many agricultural textbooks recognize farm types or types of farm system described usually by a 'typical' sample, for example, the sample British farming systems described in Watson and More's *Agriculture* (1962). Simple area criteria have been used to distinguish types. Ooi Jin Bee (1959) and Edwards (1961) have used the figure of 10 hectares as the upper limit for peasant farms. Figures of this kind are difficult to apply, however, for cash-cropping involving the use of hired labour and the planned production of several commercial enterprises does appear in areas of supposed peasant farming on holdings even below 4 hectares in size and is virtually indistinguishable from acknowledged non-peasant systems elsewhere. Although we may distinguish one farm from another as different in type, it may become impossible to draw a line between our types in the universe of farms. Some of the problems of typing farms arise from the fallacy of reasoning from the specific to the general, of contact with just a few farms whose 'types' or 'models' are then projected to include all farms (Haystead and Fite 1955, 3). The kinds of problem are well illustrated by the confusions and overlaps of early classifications such as that of Derwent Whittlesey (1936) which included such broad terms as 'shifting cultivation', too general even to be a type, almost impossible to distinguish from 'rudimentary sedentary tillage' (how sedentary or how shifting is an irregular rotation or pseudo-rotation?) and of a quite different order of classification from 'intensive subsistence tillage with rice dominant'. In the last case economic and crop criteria have been introduced. But what are we to make of 'Mediterranean agriculture' in the same list, based apparently on the assumption that a distinctive type of climate has created a distinctive association of crops and livestock? The introduction of some factor as a criterion to distinguish type, especially a physical factor, has been a temptation which few geographers, unfortunately, have been able to resist. Thus Stamp in a classification which included crofting, dairy farming, feeding and crop farming, introduced 'downland farming' as a 'somewhat specialized form of arable farming' (Stamp 1946, 62–3).

One of the earliest attempts to classify farms on a national scale was made by the Board of Agriculture in 1908. It distinguished 'wholly arable', 'wholly pasture' and 'mixed' (Napolitan and Brown 1963). In 1925 the Ministry of Agriculture distinguished the same three categories on an area basis, selecting the figure of 70% as the minimal proportion of area in arable land or pasture respectively to qualify for classification in the first two types. In 1939 the same classes were again distinguished but the critical proportion was raised to 75% and there was some subdivision of classes. In Scotland similar criteria were used for a division in 1927 into hill sheep, dairy, horticultural, poultry, pig and other full-time farms (Baines 1968), but in 1952 the Department of Agriculture for Scotland employed standard man-days as a criterion, and these were used also by Langley and Luxton (1958). Acreage have in any case always been awkward guides to farm types. As Harvey (1963) showed hop farmers in Kent frequently had only 10–15% of their land under hops. The rest was a 'manure factory' to maintain hop production. Thus only 15% of the area dominated the rest. Scola (1952) classified farms by working hours required, types of activity and acreage proportions. Bennett-Jones (1954) produced a classification combining acreage proportions and livestock numbers per 100 acres. In these two cases the combination of two or more measures was awkward. Standard man-day classifications were developed by Ashton and Cracknell (1960–1) and Napolitan and Brown (1963) and have been used by the Ministry of Agriculture. The basis of the first British types of farming map (Land Utilization Survey, 1941) is by no means clear (Board 1963), but in 1968 a colour-printed map was produced based on a one-sixth sample located by 10-km grid squares and using standard man-days (Church *et al.* 1968). The following year separate maps were produced on the same basis for each farm type. These used simple subjectively chosen proportions to determine the classes of intensity for each type (fig. 9.1) and have been made possible, like the 1968 map, by the use of a computer. Pleas have been made for isopleth maps using ratios of numbers of farms in each square deriving a given proportion of their gross output from a given enterprise (Board 1963; Napolitan and Brown 1963). Such maps are based on the geological facies maps of Krumbein (1955 and 1956) which are held to have solved similar problems of presentation. Standard man-days are far from the perfect solution to the problem of providing a good measure of the farm firm. Actual man-days which would be extremely difficult to obtain might provide an input measure, but standard man-days assume for each farm a common standard of efficiency whatever the size of unit or the capital available for expenditure on equipment. Not all of the more recent classifications use standard man-days. Barnes and Jeffery (1964) employ a mixture of input and output data together with data for natural resources and social characteristics. The analysis is complex and uses subjective cut-off criteria such as a minimum requirement of 10% area in given categories. Proposals for a world-wide typology of agriculture appear to be based on three main categories of criteria, each with considerable subdivision:

1. social and ownership features;
2. organizational and technical features;
3. economic (productive features).

Choice and relative weightings will be highly subjective. Measures will vary greatly in quality. Despite this Kostrowicki claims that the solution of the problem of 'exact methods of the integration' of these criteria must be solved if the typology is to be 'certain and objective' (Kostrowicki 1964*a* and *b* and 1966). By the end of 1970, however, the methodological problems for creating a world agricultural typology had not been solved, although some agreement on the criteria to be employed had been reached (Kostrowicki 1970), and a search was still being made for the most suitable method of combining indices representing various agricultural characteristics and for clearer definitions of farming systems and related systems of society, land tenure and land utilization (Kostrowicki 1969). We may, however, wonder whether the pursuit of a world 'omnibus' classification can either be of much scientific use (Simpson 1965), or can provide a meaningful basis for the solution of agricultural problems through comparative studies.

Classification of farming areas

Most classificatory work by agricultural geographers has not been of farms but of the areas for which agricultural data have been published. The problems of adequately locating and interpreting such information have already been discussed. The nature of the problem has been illustrated by Scola (1952) who cited Fife, classified as 'arable with livestock feeding', but containing at the time of classification: 351 farms which had cropping with livestock, 277 dairy farms, 177 stock-rearing and feeding farms, and 81 stock-rearing farms. Such classification can in no sense be regarded as a farming classification. It refers solely to the enterprise data which happen to be located for the purpose of publication within a particular administrative unit known as Fife. However, despite the problems of interpretation, such data should not be neglected and attempts to classify the areas concerned by the association of enterprises recorded gives at least some idea of relative local importance according to the measures applied.

The problem of deciding on the extent of relative local importance preoccupied J. C. Weaver (1954*a* and *b*; 1956). He proposed a method of analysis superior to simple inspection of the relative weights of the enterprises in a given area. This was to apply the standard deviation formula:

$$\sigma = \sqrt{\frac{\Sigma d^2}{n}}$$

where d = the difference between the actual proportion assigned to an enterprise in a given census unit and the theoretical proportion were a given crop association to be adopted. By this method Weaver sought to find the number of enterprises that minimized the difference between the actual and theoretical enterprise combinations. Weaver restricted his calculations so that in testing for single-crop dominance only the proportion of the largest crop area was used, for two-crop dominance only the proportions of the two largest crops, and so on. This method reduces labour but it can lead to errors of classification in a few mar-

TABLE 9.7 *A worked example of the application of the standard deviation formula to enterprise combination classification*

Suppose the percentages of a given area under crops are:

	Wheat	*Barley*	*Potatoes*	*Sugar Beet*	*Beans*
	49	26	14	7	4
Then for one-crop dominance we test:					
Theoretical combination (%)	100	0	0	0	0
Differences (d)	51	26	14	7	4
d^2	2601	676	196	49	16
Σd^2			3538		
$\Sigma d^2/n$ (=number of crops)			3538		
Calculation of the square root is unnecessary					
For a two-crop combination:					
Theoretical combination (%)	50	50	0	0	0
d	1	24	14	7	4
d^2	1	576	196	49	16
Σd^2			848		
$\Sigma d^2/n$			424		
For a three-crop combination:					
Theoretical combination (%)	33	33	33	0	0
d	15	7	19	7	4
d^2	249	53	373	49	16
Σd^2			738		
$\Sigma d^2/n$			246		
For a four-crop combination:					
Theoretical combination (%)	25	25	25	25	0
d	24	1	11	18	4
d^2	576	1	121	324	16
Σd^2			1038		
$\Sigma d^2/n$			260		
For a five-crop combination:					
Theoretical combination (%)	20	20	20	20	20
d	29	6	6	13	16
d^2	841	36	36	169	256
Σd^2			1338		
$\Sigma d^2/n$			268		

Thus the three-crop combination provides the lowest deviation and may therefore be regarded as the best fit.

ginal cases. The full procedure as described by Thomas (1963) and Coppock (1964*b*) is outlined in table 9.7.

It cannot be too strongly emphasized, however, that all that has been achieved is a statistical test of crop dominance. The best fit solution is the best solution for classification in the absence of other information. Simple areal dominance or even production dominance does not necessarily tell us much about the importance of enterprises in the system. In our calculation sugar beet and beans added together cover 11% of the cropped area. The sugar beet may be extremely profitable. The beans may provide a break crop. The percentages may in any case reflect a changing, not an equilibrium, situation and 'dominance' may be meaningless in organizational terms. The method has been used to show crop, livestock and enterprise combinations by National Agricultural Advisory Service Districts in England and Wales in 1958, employing man-days instead of area units (Coppock 1964*a*). A major problem was the allocation of man-days to enterprises. Information at farm level was not available and a simple broad classification of enterprises had to be adopted. Wheat, barley, sugar beet and potatoes were recognized as cash crops. The man-days for other crops and grass were allocated to grazing livestock (sheep and cattle) in proportion to their feed requirements expressed as livestock units. In the case of barley, however, approximately three-quarters of the crop is for feed grain and a large part of this is kept back on the farms. The distinction between beef and dairy cattle was also difficult to make and involved again an arbitrary procedure, while the separation of horticulture and cropping involved the allocation to horticulture of coarse vegetables grown as field crops, and the creation apparently of the extra enterprise of horticulture in predominantly cropping areas.

Land classification

The classification of land according to its quality for agricultural use is still regarded as important as land quality should influence the selection of enterprises. Numerous methods of assessing land quality have been devised, but many have been unsuccessful, particularly where it has been assumed that 'inherent fertility', as measured by the physical and chemical characteristics of land, could be meaningfully demonstrated without reference to social and economic considerations. Nevertheless, the governments of those nations that wish to develop their land resources or whose policy it is to conserve good-quality agricultural land have persevered in their attempts as they see a practical value in the classification of land.

The land capability approach, in which land was graded according to its susceptibility to soil erosion under differing farming and forestry systems of production, has been successfully applied to much of the developing world, and even to a part of Shropshire (Mackney and Burnham 1964). The greatest practical difficulty is that of distinguishing between land classes. This point may be illustrated by reference to the Agricultural Land Classification Map of England and Wales being prepared by the Agricultural Land Service (1968). Agricultural land is graded according to the degree to which its characteristics impose long-term limitations on agricultural use, particularly on yields. The chief physical factors taken into account are rainfall, transpiration, temperature, exposure, slope, soil wetness, soil depth, soil structure, stoniness and available water capacity. Five grades are distinguished, but again there are highly subjective factors in the classification which make interpretation difficult. For example how precise can we be about Grade III described as:

'Grade III: Land with moderate limitations due to soil, relief or climate, or some combination of these factors which restrict the choice of crops, timing of cultivations, or level of yield. Soil defects may be of structure, texture, drainage, depth, stoniness or water holding capacity. Other defects, such as altitude, slope or rainfall, may also be limiting factors; for example, land over 400 ft which has more than 40 in annual rainfall (45 in in North West England, Western Wales and the West Country) or land with a high proportion of moderately steep slopes (1 in 8 to 1 in 5) will generally not be graded above III.' (p. 3)

One is hardly surprised to discover that the grades may merge despite sharp lines on the map, that complicated patterns have been avoided and predominating grades applied to whole areas, or that land may be assigned to a grade for 'widely different reasons' (p. 5).

The economic instead of the physical classification of the land offers the approach of tackling the problem through an analysis of the land-use pattern or the farm business. A summary of studies in the Far East, in which the emphasis in classification is on the intensity of land use, is provided by Lewis (1969). He claims that research work in the Far East shows a relationship between 'land productivity', wages and returns to capital, and that land classification is an essential prerequisite for a theory of rent. This supports conclusions reached by economists in North America who note that farm income is closely related to land quality and that land classifications can be used as means of predicting farm income expectancy (Conklin, 1959).

Dudley Stamp (1948) is still the best-known exponent of the value of land-use data for land classification purposes. He assumed that under stable economic and social conditions agricultural activity would adapt itself so as to make the best use of land resources. The most demanding systems of production would therefore concentrate on the best land and the most extensive on the poorest. Coppock (1962) has added a time dimension to this notion by recording that Britain's best and worst agricultural land has not changed in its general use during the industry's economic fluctuations of the last century, while land of intermediate value has done so frequently. The value of Stamp's assumption lies in its simplicity, but as an extreme generalization it only reveals gross differences in land quality, while at the individual farm level it makes the untenable assumption that all farmers have equal knowledge, ability and information and would therefore make the same use of the same resources. In later work, Stamp tried to measure the productive capacity of land. He discarded land values as influenced by many factors, and cash values as subject to change. Instead he adopted a measure of yield defined as 'the potential production of one acre (0·4 hectare) of good average farmland under good management'. His 'very rough equation' was as follows (Stamp 1960, 113–14):

0·4 hectare of First-class Land = 2 PPU (Potential Production Unit)
0·4 hectare of Good General Farmland = 1 PPU
0·4 hectare of Medium Light or Medium General Farmland = 0·5 PPU
0·4 hectare of Poor Land = 0·1 PPU.

Again, however, there are problems of subjective yardsticks and of management difficulties which frequently appear to be greater than supposed differences between either different systems of production or soil types. A similar criticism can be made of a more recent attempt to classify agricultural land in south-east England (Hilton 1962; 1968). Two sources of farm data were used to classify farms. Firstly, yields from all crops covering more than 10% of the farm were compared with national averages and given a grading between 1 and 5. The mean value of all the enterprise gradings was then used to classify the farm. Secondly, this grading was adjusted according to the farm's 'institutional' characteristics. This principally meant size, the farm being upgraded one class if over 120 hectares in extent and downgraded if less than 20 hectares. No attempt was made to ascertain whether land or management factors were responsible for varying crop-yield levels and, as the function of the classification was to assist short-term planning decisions, this problem, it was claimed, did not need to be resolved (Hilton 1968, 140).

Quantitative techniques in agricultural classification

Reference has already been made above to classificatory systems using measurements instead of presence or absence of selected criteria. These systems have either used quite arbitrary methods of grouping their data or have tried 'best fit' solutions. Other methods involve determining the between-group variance either by ranking the data and observing 'steps' in the rank-order or by constructing matrices listing the measurements against the criteria and seeking groups. Such groups may, however, be difficult to define and measures of similarity, association and correlation have been devised to compare the data (Harvey 1969, 338–48). Principal components and factor analysis have also been used for classificatory purposes in order to discover the relationships involved (principal components) or to test whether a theory fits the data (factor analysis). Important contributions to the study of agricultural geography using these methods have been made by Kendall (1939), Henshall and King (1966), Aitchison (1970) and Munton and Norris (1969). Accounts of the differences between the methods and the techniques required may be found in Cattell 1965; Cooley and Lohnes 1962; Kendall 1957 and Rummel 1968. Important though these methods are, they are not a universal solvent for problems of classification or of the factors governing location. Correlations are not relationships and the degree of correlation even where relationships are known to exist may not be a reliable indicator of the relative importance of the relationships. In any research situation data may overlap, be omitted, unknown, not measurable or measurable only in a way which introduces bias into the results. To employ these techniques successfully a great deal has to be known with regard to the nature of the relationships involved.

10 Patterns of distribution

Distribution patterns can rarely be viewed objectively. In the recognition of group patterns the problem of the subjective element is especially acute and may involve the linking of unlike features such as crops and livestock, difficult to assess by any common measure, and often involving a problem of which feature to choose first in creating regional or classificatory hierarchies (Weaver *et al.* 1956). This is not to suggest that such features are not located independently of the observer, but that the recognition of patterns and of areal co-variation depends to a large extent on the methods of measurement and the type of classification chosen. This is true even where quantitative methods are used, for such methods help only by making definitions more precise and comparisons clearer. Regionalization and classification are interrelated processes of thought. They are not, however, the same process. Properly speaking we begin with classification for our concern is with the agricultural elements, and, as Grigg has remarked, 'areas are not discrete objects as stones or plants' and 'taxonomists do not consider position, or location, when classifying their individuals' (Grigg 1966). We may therefore divide enterprises or farms into classes on the basis of certain differentiating characteristics. When, however, we plot our farm classes or enterprise classes on maps with appropriate symbols we should define 'regions' only where similar symbols are contiguous. 'It is this step in the procedure which makes regionalization so different from classification as a taxonomic procedure' (Grigg 1966). In practice agricultural regions have been defined on the basis of local 'dominance' of given enterprise or farm type, and frequently by the use of very small samples. They often mask, therefore, a considerable variety of agricultural practice which may be extremely important in any attempt to understand agricultural patterns.

The agricultural region

Many geographers have supposed that actual 'integrations' or 'syntheses' of geographic phenomena existed in reality, or at least complexes 'of loosely interrelated segments, each of which is formed of closely interrelated phenomena, together with some phenomena which show little or no relation with the others' (Hartshorne 1959, 114).

Given the tradition that the concept of such 'element-complexes' or 'regions' is of fundamental importance in geographical studies (Symons 1967), it is hardly surprising that most agricultural geographers hitherto have been concerned mainly with problems of regional definition usually of a general character, rather than with the solution of problems of agricultural location. In consequence little of their work has been of use for the solution of farming problems, except in so far as it has added to our descriptive knowledge of agricultural distributions. However, many of the problems of definition involved have arisen more from the nature of the classifications employed than from the nature of the farming systems observed.

Despite our criticisms we need to deal with the concept of the agricultural region because it has been used in the past, and is still being used by geographers and economists who take a different view from our own. Most agricultural geographers today would reject the notion of using 'natural regions' as agricultural regions in the manner of Roxby (1925), even though an environmental approach of this kind is still not unknown and is at least suggested by the traditional 'physical basis' approach of some of our texts (see above, p. 14 and criticism in Gregor 1970, 114). The agricultural region, if it is to be defined at all, should be defined in agricultural terms, that is by crop, livestock, or enterprise association data, or by measurements of farming processes or of farming organisation (Buchanan 1959, 5). The trouble with much agricultural geography is that it is a small step from the mapping of such data and the recognition of areas of 'dominance' by a particular enterprise, association of enterprises or farming type, to the recognition of such areas as distinctive regions 'readily recognizable on the ground' (Symons 1967, 196), dominated by an enterprise possessing 'a sort of natural monopoly' (Baker 1925), and possessing a distinctive landscape including patterns of land-use, field systems and settlement patterns (Simpson 1957; Grigg 1969). Eventually it is the 'belt', the 'zone', the 'region', or even the 'economic landscape' (Lösch 1954, 135–7, 216) that has an existence of its own. The region becomes an area 'with a high degree of homogeneity of one or more criteria' (Spencer and Horvath 1963) and not an area simply occupied by certain features and defined by certain arbitrarily chosen criteria. This is not so much better than Russell Smith's definition of the Corn Belt as 'Nature's conspiracy to make man grown corn' (Russell Smith 1925, 290, quoted in Spencer and Horvath 1963). Spencer and Horvath avoid Smith's environmentalism only to fall into the trap of viewing space as a series of containers, frequently exclusive, following the Kantian approach that the whole of reality may be divided into sections in terms of either space or time (Harvey 1969, 207–8). Although our approach to agriculture is areal or locational, and our methods deal mainly with areal or locational definitions and measurement, nevertheless we are dealing essentially not with areas but with the elements of agricultural systems.

Single enterprise distribution: areas of concentration

All enterprise 'regions' are areas with a recognizable degree of concentration of crops or livestock (for a study of concentration see, for example, Britton and Ingersent 1964). We may think of the enterprise concerned as consisting of a population defined as production or areal units or even defined in terms of values. We must, however, be careful to distinguish between data in an absolute form which may be treated as we would treat a population and data in relative form in which the relationships refer to other enterprises, proportions of holdings (by inputs, areas, or values), or proportions of different kinds of land which may be affected by other enterprises (such as arable or tillage land). The distribution pattern of wheat, for example, is quite different when the data refer to cropland from the patterns derived from data based on arable land, tillage land, or all land. Although enterprises may in effect be treated as population, and livestock enterprises frequently are, crops are hardly ever described in density terms (i.e. not just yield but returns from the total farm, parish, or county area, for example) or in terms of nearest-neighbourhood analysis (Clark and Evans 1954) which might be used to examine the distances between the fields occupied by a given crop. The main method of quantifying general distributions so far used has been by a location quotient or by a coefficient of localization.

A LOCATION QUOTIENT (Florence 1948) compares a region's share of some enterprise with its share of some base aggregate. This technique was originally devised for the study of industrial location. The simple formula is:

$$\frac{E_C \, . \, T_N}{E_N \, . \, T_C} \quad . \; . \; . \; . \; . \; . \; . \; . \quad 1$$

where E_C is the area of the enterprise in the county or state, E_N is the national area of the enterprise, T_C is the total area of specified enterprises in the county or state, and T_N is the national area of specified enterprises. Simplicity of calculation and comparison makes the quotient useful in the early stages of research. However, it can be difficult to interpret and has been claimed as meaningless when used on its own (Isard 1960, 125–6).

A COEFFICIENT OF LOCALIZATION (Florence 1944; Chisholm 1962, 93) is a measure of relative regional concentration in which comparison is made not between regions but between enterprises by calculating the mean differences between local and national proportions of the area occupied by given enterprises. The difference for each county or state is calculated by:

$$\frac{E_C}{E_N} - \frac{T_C}{T_N} \quad . \; . \; . \; . \; . \; . \; . \; . \quad 2$$

The resultant differences are either positive or negative. Normally the positive differences are added together and divided by 100 giving a coefficient ranging between 0 (enterprise distributed exactly as cropped area) and 1 (enterprise highly concentrated). Negative differences can also be added together and divided by 100 to give an index of relative absence of the enterprise. Again there are problems of interpretation for only the area of relative concentration (or of relative absence) is assessed.

However crude and subjective these methods may be we can begin to distinguish types of single enterprise areas of concentration. First we may distinguish crops or livestock which are highly localized from those which are more widespread in their distribution although they may still exhibit within their distribution pattern areas of marked concentration. The more highly localized crops despite recent trends still include most vegetables and fruits, more especially those with a small production and limited market and which require special environmental conditions, special skills or very high levels of capital investment. In temperate countries such horticultural crops as celery, bulb flowers and tomatoes grown in the open have special environmental requirements, as have early potatoes and other fruits and vegetables grown to take advantage of seasonal high prices. The use of glass and other techniques to promote earliness is a technical innovation which could encourage wider distribution, but can be offset by cheaper transport and easier international trading conditions. Thus Britain's entry into the E.E.C. could have an adverse effect on the production of fruits and flowers under glass in south-east England. Minimization of transport costs could result in the concentration of most 'earlies' in southern Europe at locations convenient for conveyance by road, rail, or air. Brassicas tend to concentrate around towns, and some soft fruits around jam factories. However, the locations of some highly localized crops are as yet difficult to explain such as the concentration of rhubarb around Leeds. Frequently it appears to be the case that of a wide range of possible locations one area establishes an early reputation and this is strengthened and confirmed by the subsequent development of marketing or processing arrangements with accompanying external economies of scale, which then give the area an advantage over all possible competitors. This is certainly the case in Britain with a number of crops such as tomatoes under glass, sugar beet and hops. There may be special physical advantages, but these advantages are in effect only perceived and utilized by economic systems, so that the reasons for the development of a given production in a given area are economic. Once established inertia helps to maintain an enterprise for 'the specialized area . . . will fight to the last ditch rather than accept a change of specialization' (Buchanan 1959, quoted in Harvey 1963). Such 'regionalizations' of enterprises will tend to increase as the intensity of production increases unless markets are considerably enlarged.[1] In the case of livestock enterprises high degrees of localization are less marked, but attention should be drawn to the extraordinary concentration of beef production in eastern Scotland, dairying in Holland and pig production in Denmark, and also to the localization of special products such as famous cheeses or of particular breeds of livestock. Within broader distributions special attention should be paid to local concentrations of production such as dairying in Cheshire, peanuts in Alabama and Georgia, or wheat in the Paris basin. High degrees of localization of crop production are notable in the tropical world more especially among export crops.

In examining types of concentration we may distinguish crops with a single nucleus of concentration from multiple concentration crops and from crops with multiple concentration plus a more widespread low density distribution. We may also distinguish concentrations with no particular centre from concentrations with centres or core areas. In addition it is especially important to examine the changes in the patterns of concentration which take place over time. Some patterns are relatively constant, even maintaining the same core area over a long period and exhibiting only expansion and contraction about the core. Others show expansion and contraction within the same general location but exhibit some shifting of core. Others again appear to move across a country and if exhibiting any constancy at all exhibit only that of direction.

Innovation and diffusion

The creation of distribution patterns is largely the result of two processes:

(*a*) innovation and its accompanying process of adoption;
(*b*) diffusion or spread, that is the direction and rate of adoption of innovation between persons and locations.

Metcalf (1969, 65) distinguishes three kinds of agricultural innovation: those which leave the final product unchanged but reduce the unit cost of the output, tending to result in increased total output; those which provide new products or services; and those which change business organization or marketing structures. In recent years the spread of a new farming technology has been remarkably rapid, but it has

1. For discussion of the increasing regionalization of British agriculture see Coppock 1971.

consisted of the diffusion not of a single group of ideas but rather of a variety of innovations one after the other requiring constant adaptation on the part of the farmer. The general effect has been not to reduce regional contrasts in distribution patterns through the exercise of a greater control over the environment, but to create in most locations more marked concentrations of particular forms of production and to eliminate some of the more specialized enterprises from locations of less importance. The contrast in productive and earning capacity between the successful adopter of innovations and the farmer who still cultivates by traditional methods becomes greater year by year. Innovations have affected all the enterprises and apparatus of farming. Their adoption has tended to be more rapid among variable than fixed inputs. Desirable though new buildings or a new combine harvester may be, farmers find it easier to try new seeds, livestock feed, or fertilizer. Moreover, innovation in fixed inputs, often demanding a complete reorganization of the production system is likely to involve considerable demands on capital, necessitating a change in the financial situation of the farm and a greater element of risk.

Description of diffusion and the related study of origins are old established parts of agricultural history and historical botany (Vavilov 1949–50; de Candolle 1883). They are also an important feature of historical geography and of more speculative work on the origins of agriculture and settlement such as Sauer's, *Agricultural Origins and Dispersals* (1952). More recently they have begun to play a role in the study of the evolution of 'regions' or of distribution patterns as in Spencer and Horvath's 'How Does an Agricultural Region Originate?' (1963). More detailed studies of particular diffusions have been attempted by Ryan and Gross (1943) and Bowden (1965), factors affecting the speed of diffusion have been studied by Jones (1963), and the field has been generally reviewed by Henshall (1967), Jones (1967) and Metcalf (1969). Special studies of innovation adoption have been made in the underdeveloped world, notably by Schultz (1964). Theoretical work applicable to all kinds of diffusion is not of special concern here, but some of it has been tested in part by agricultural examples, notably in studies by Hägerstrand (1953), Brown (1965) and Harvey (1966). An important feature for comment is the recognition by some workers of stages in adoption (Rogers 1958 and 1962) and by others notably Hägerstrand (1952) of stages in the diffusion process. Stages are an important aid to thinking where there is evidence of changes in diffusion patterns and processes. Where they lead to the notion of growth and of a regional spread as possessing organic qualities they can, however, promote misunderstanding.

Examples of single enterprise distribution and diffusion patterns

SINGLE CONCENTRATION WITH PERSISTENT CORE: HOP FARMING IN KENT (Harvey 1963)

Hop cultivation in England expanded its acreage by a series of long-term fluctuations mainly between 1810 and 1885, after which the acreage contracted. In Kent there were two districts, central and east Kent,

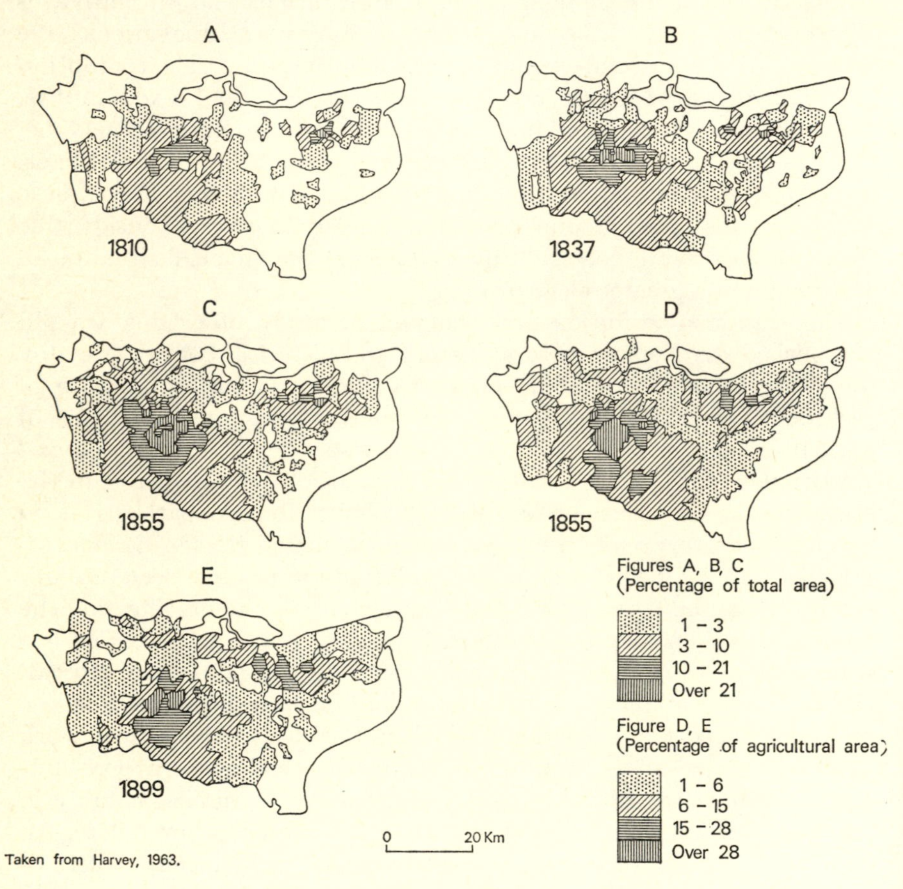

10.1 *Density of hop cultivation in Kent for selected dates in the nineteenth century.*

of which the latter was less important and distinct because it concentrated on a high-quality product for only one sector of the market. Harvey's survey of changes in the distribution of hop farming in Kent showed (fig. 10.1) the remarkable persistence of a core area with a marked tendency for the density of hop cultivation to decline with distance from the core (fig. 10.2). This phenomenon appeared to have 'little respect for differing soil conditions' despite the fact that many studies have suggested a close relationship between hop yields and soil drainage and texture. The statistical 'centre of gravity', located throughout the period of expansion between Wateringbury and Yald-

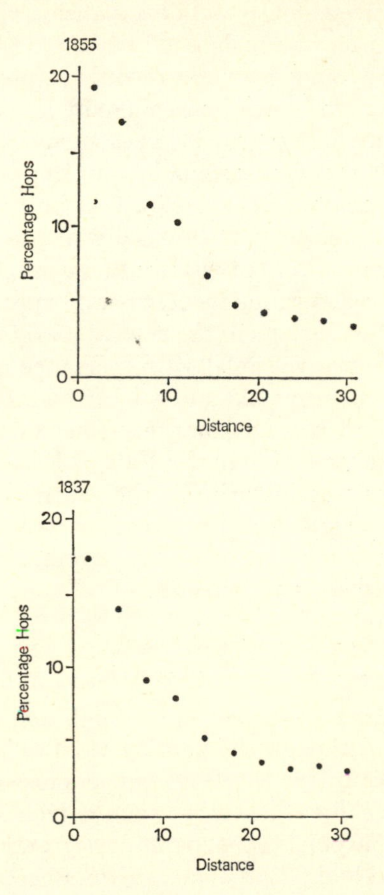

10.2 *Decline of density of hop cultivation with kilometres from Wateringbury.*

ing, persisted during the period of contraction from 1885 to 1899, the percentage decline becoming more marked with distance from the centre. Indeed the relationship between distance and density was 'far closer' during the period of contraction reflecting that increasing degree of agglomeration and regional specialization which Chisholm suggests have been developing at all scales in the economically more advanced nations (Chisholm 1962, 195–6). Hops are still highly localized in the same area of concentration (Best and Gasson 1966, 42–3 and 81), although some of the factors (e.g. availability of labour, local manure sources) have changed, and production is being concentrated into larger units. Harvey showed that the concentration of production in the nineteenth century was caused by the operation of two specific processes – agglomeration and cumulative change – while expansion was caused by diminishing financial returns in the face of expanding demand. The main agglomeration factors among several recorded were marketing and the development of a parish 'mark', easier credit in established locations, longer leases in the main hop-growing area encouraging persistence and the development of a regular seasonal labour migration from London to the area. No doubt contact between farmers and the dissemination of information played a major part especially in the early spread of the industry. This particular kind of distribution can be analysed by means of a gravity model to determine its centre and is particularly useful for the observation of fluctuations in the spatial margins of production and for the understanding of the relative importance of certain locational factors. There are numerous examples elsewhere of products that are mostly located around a single market focus or around a group of markets, e.g. market gardening in the Vale of Evesham focusing on the Evesham core which was historically the centre of origin, the burley tobacco area of East Tennessee with its core at Greeneville or the Bright Tobacco Belt of the Piedmont of North Carolina with its markets at Winston-Salem, Durham and Raleigh.

SINGLE CONCENTRATION WITHOUT A CORE: DAIRY FARMING IN SOUTH CHESHIRE (Simpson 1957; Mercer 1963)

South Cheshire has been described as the 'high temple of dairy farming' (Mercer 1963, 180). Although the density of cows is no greater than in other parts of the country, the local dependence on dairy farming is considerable with few livestock other than cattle and over 80% of the farmland in grass. The chief crops, oats, barley and kale are for fodder, although some wheat and potatoes are grown. Sheep, pigs and fowls are also kept, but are much less important than cattle, and except on specialist holdings are tending to decline in importance as the advantage of specialization becomes greater. Approximately half the holdings are under 20 hectares. The small size of farm business has helped to encourage persistence in dairy farming and the development of a tradition, because although incomes have been low they have also been reliable. Simpson shows the persistence of the South Cheshire dairying area in the face of changing technology and economic conditions. Cheese making has declined and the farms have turned more to liquid milk production to supply the rapidly expanding urban markets. No core area emerges in Simpson's study, but rather a grouping of areas with rela-

10.3 *Some patterns of distribution of features related to dairy farming in south Cheshire.*

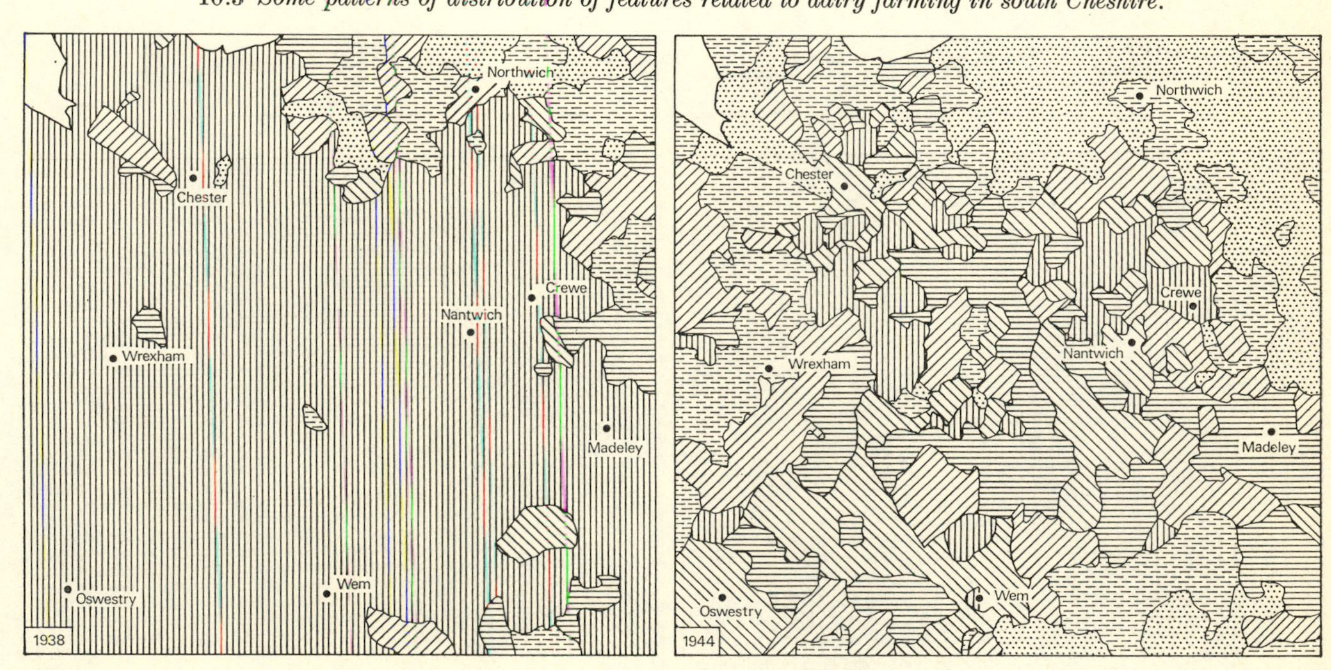

The percentage of cultivated land in permanent grass in (a) 1938 and in (b) 1944

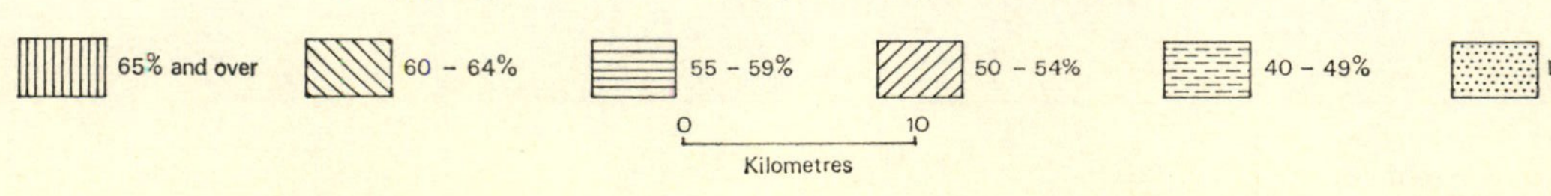

10.3 (*continued*)

Cows and heifers in milk and cows in calf per forty hectares cultivated land in (a) 1938 and in (b) 1948

10.3 (*continued*)

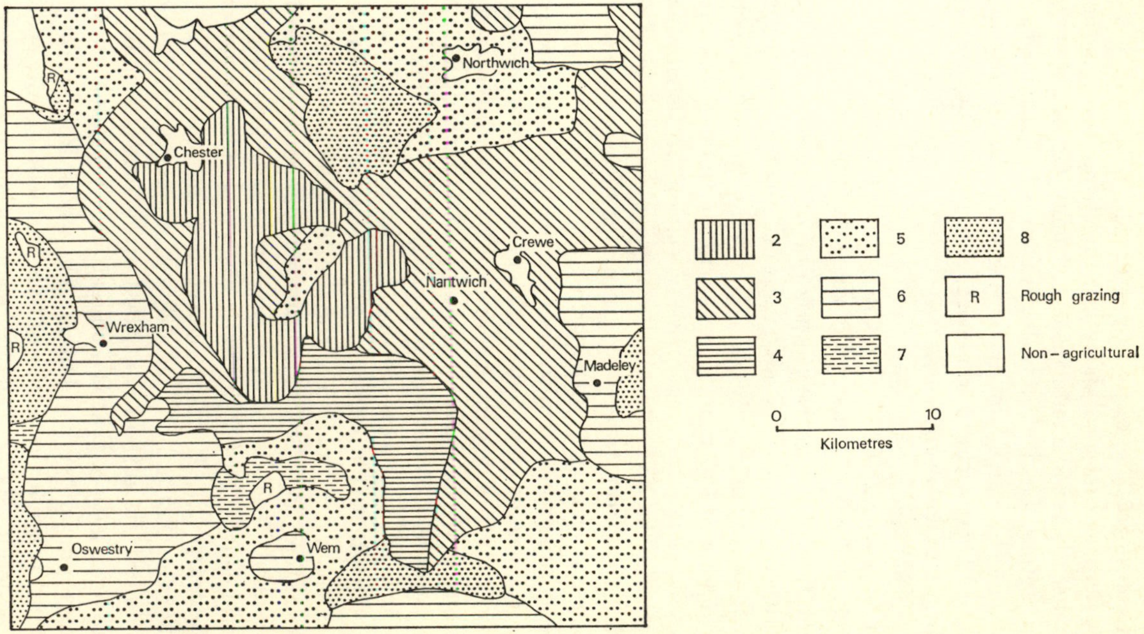

Northwich
Chester
Crewe
Nantwich
Wrexham
Madeley
Oswestry
Wem
R

2
3
4
5
6
7
8
R Rough grazing
Non-agricultural

0 10
Kilometres

Pasture groups Based on Simpson 1957

tively high densities of dairy cows, the centre of which is the hill country immediately north of Whitchurch. There is only a weak association between the heavier soils, the 'better' grasslands or pasture groups and the main concentrations of dairy cattle (fig. 10.3), and it would seem that the existence of a local dairy-farming tradition plus the difficulty of developing alternative enterprises on small farms are the important locational factors. The lack of a single core is in part due to the size of the area and the spread of marketing facilities. There is no single organizing centre or focus of transport, only a common interest by most farmers in a particular enterprise.

MULTIPLE CONCENTRATIONS OR ENTERPRISE 'BELT'; COTTON FARMING IN THE UNITED STATES OF AMERICA

The recognition of enterprise belts is one of the oldest features of the attempt to discern distribution patterns. Most such recognition has taken account only or mainly of the general character of such distributions and has failed to pay much attention to the variable nature of the factors involved nor to the marked variation in distribution pattern existing within the belt. Most belts are agglomerations of distinct or nearly distinct concentrations or groupings of a particular enterprise. Often each concentration, although of the same enterprise in general terms, e.g. cotton, wheat, maize, is of a different enterprise in marketing terms, i.e. consists of a particular variety of the enterprise, is produced for a particular set of dealers or markets, or is produced at a particular time of the year. Given the early predilection of agricultural geographers for a physical environmental explanation of the phenomena they observed, it is hardly surprising that the limits of such crop belts were found to coincide better with general physical isopleths than with anything else. Baker's distinction of the limits of the Cotton Belt by determining its extent and then searching for coincident physical isopleths provides a firmly established example. Some attention was paid to belts with several centres by Lösch (1954, 87) who noted the cotton districts around New Orleans, Houston, Galveston and the smaller ports, but argued that the Cotton Belt was still a unit specializing in one product. It is the view of the authors that the Cotton Belt is a 'unit' only at the level of an aggregation, that is a collection or plurality whose members are in geographic proximity but among whom interaction is small or non-existent. It is not the region but the farmers who are specializing in cotton production. Many of them produce other crops or rear livestock, and sometimes attach more importance to these other enterprises. The distribution patterns of these other enterprises could be regarded as 'parts' of other belts which thus overlap with the cotton belt. To adopt dominance criteria in order to assign exclusive locations to belts may achieve geographical neatness and lead to correlations with physical isopleths. It may also prevent us from understanding the nature of the distributions involved.

There are twelve major areas of concentration of cotton production (fig. 10.4). These have been related to climatic and soil factors. Important though these factors are the operation of others is suggested by the variation in areas of cotton concentration at different dates (Gray 1941; Erickson 1948; Haystead and Fite 1955; Griffin *et al.* 1963; Prunty 1951; White *et al.* 1964). The other factors include:

(*a*) accessibility to market and transport – denied in Perloff *et al.* (1960, 358) but certainly important at some stages of development (Gray 1941, 904 and 907);

(*b*) the early tie between producers and ports due in part to the need for capital provided chiefly by the entrepreneurs of Charleston, Savannah, Mobile and New Orleans;

(*c*) the early substitution of cotton for indigo in South Carolina and the differing possibilities for substitution elsewhere, e.g. livestock in South Carolina, subsistence food crops in the south-west, wheat in the Carolinas and sugar in Louisiana (Gray 1941, 917);

(*d*) The changing roles of cotton in the agricultural economy from its initial attraction as a garden crop requiring little capital, especially suited to small holdings in the south-east (Gray 1941, 688–9), to its development as a plantation crop favoured by booms and ready availability of capital, more especially in the Mississippi Valley and Texas;

(*e*) The link between production in the piedmont and the inner coastal plain of Georgia and the Carolinas and the development of southern textile manufactures.

Each area of concentration, developed from a different point in time, has had a different history and presents some differences of pattern, structure and relative importance in the whole. Today production in Texas and the new irrigated areas of Arizona, New Mexico and California has the largest share of the market because there the average size of cotton farm is much larger and more farms are able to take advantage of mechanization, and also because relative isolation gives greater freedom from disease.

A MIGRATORY DISTRIBUTION OR MOVING FOCI OF CONCENTRATION: COCOA IN GHANA

Although some Ghanaian farmers adopted the new cocoa crop on their existing holdings most planters sought new areas of production, preferring wherever possible heavily forested areas, and moving on to new lands whenever the production from old plantings had begun to decline. The situation is fairly typical of the development of commercial

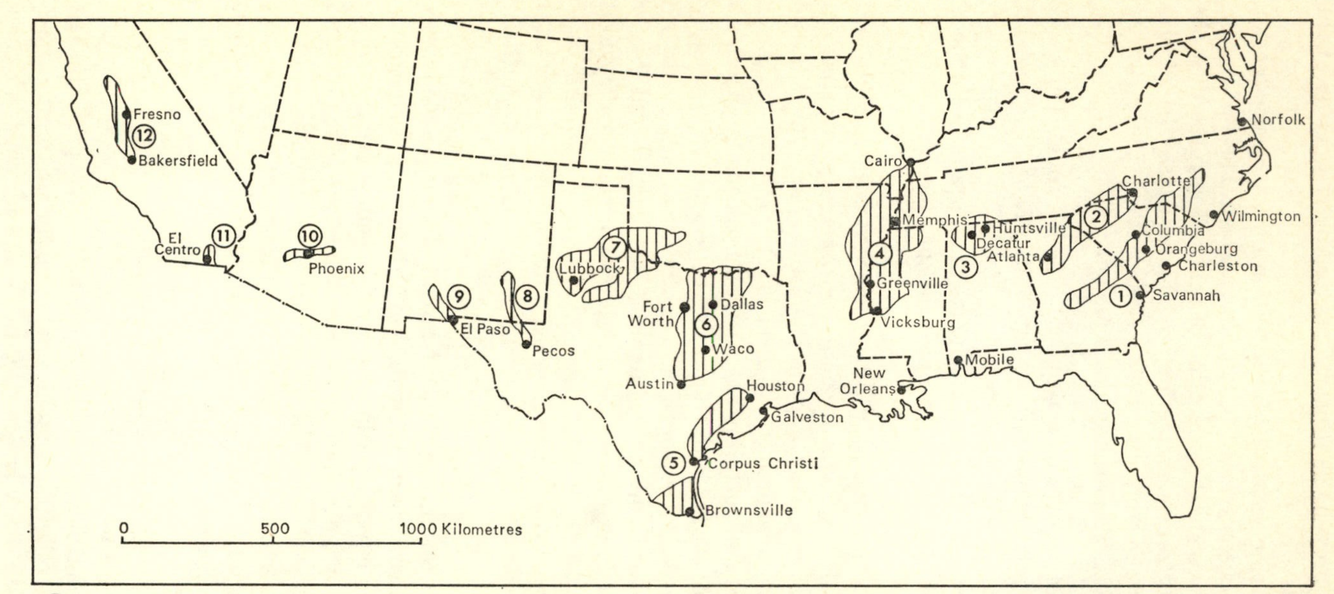

1. The inner coastal plain of Georgia and the Carolinas
2. The Georgia and Carolina piedmont
3. Northern Alabama
4. The Mississippi alluvial valley
5. The south Texas coastal plains
6. The prairies of north Texas
7. The high plains of west Texas and southwestern Oklahoma
8. The Pecos Valley of New Mexico
9. The Mesilla (Rio Grande) valley of New Mexico
10. The Salt River Valley of Arizona
11. The Imperial Valley of southern California
12. The Great Valley of California

Areas determined subjectively from inspection of distribution maps in sources quoted in text.

10.4 *Major areas of concentration on the cultivation of cotton in the U.S.A. in the 1950s.*

perennial crops in the tropics whether produced on plantations or grown by smallholders as in Ghana. Akwapim became the initial centre of cocoa planting and exporting, firstly, in established settlements such as Aburi and Mampong (fig. 10.5), and secondly, by the migration westwards from about 1892 of southern Ghanaian cocoa farmers who created camps or new hamlets and villages. (Hill 1963; Hunter 1963). This migration developed the historic core area of cocoa planting in Akwapim and across the Densu River in Akim Abuakwa. From this area the rapid spread of cocoa was favoured by:

(*a*) its increasing commercial attractiveness compared with other possible cash crops, notably coffee and the oil palm (which together with wild rubber had provided capital for the purchase of cocoa land);

(*b*) the existence of large areas of relatively unoccupied rainforest;

(*c*) the rapidly growing market for cocoa in Europe, the existence of European merchants at the ports, and of rubber traders like Peter Botchway, who also bought cocoa and invested in cocoa land (Hill 1963, 173 n.2);

(*d*) the ready adoption by cocoa planters of systems of land purchase such as the *huza* system of the Krobo (Hill 1963, 16; Hunter, 1963);

(*e*) the early development of a tradition that cocoa savings should be made wherever possible and invested in fresh purchases of cocoa land, often creating widely scattered holdings, but leading to a rapid advance of experienced growers onto new lands. The technique has been described as 'leap-frogging', the leaps becoming much greater after the advent of motor transport (Hunter 1963, 644).

The core area had the advantage of ready access to port, but expansion inland was limited by the considerable expense of head-loading and political struggles between Ashanti, the small states and the British. It depended also in part on the development of a credit system and the attraction to the cocoa areas of large numbers of labourers. Migrant labour was employed in the first stage of cocoa expansion by farmers from Akwapim who acquired widely spaced holdings too large in size for one family to operate and who still regarded the Akwapim town from which they came as home. Many Akwapim cocoa pioneers became creditor-farmers or farmer-financiers, and as cocoa planting expanded beyond the original core area from about 1900 onwards a second stage of investment expansion appeared in which labourers were employed in very large numbers. Fig. 10.5 indicates the direction of cocoa planting westwards from the core area to Kumasi by 1900 and northwards into the Volta Region (then German Togo) (Dickson 1969, 166–9). Cultivation was stimulated by railway construction, completed from Sekondi to Kumasi by 1903 and between Accra and Kumasi by 1923. The latter

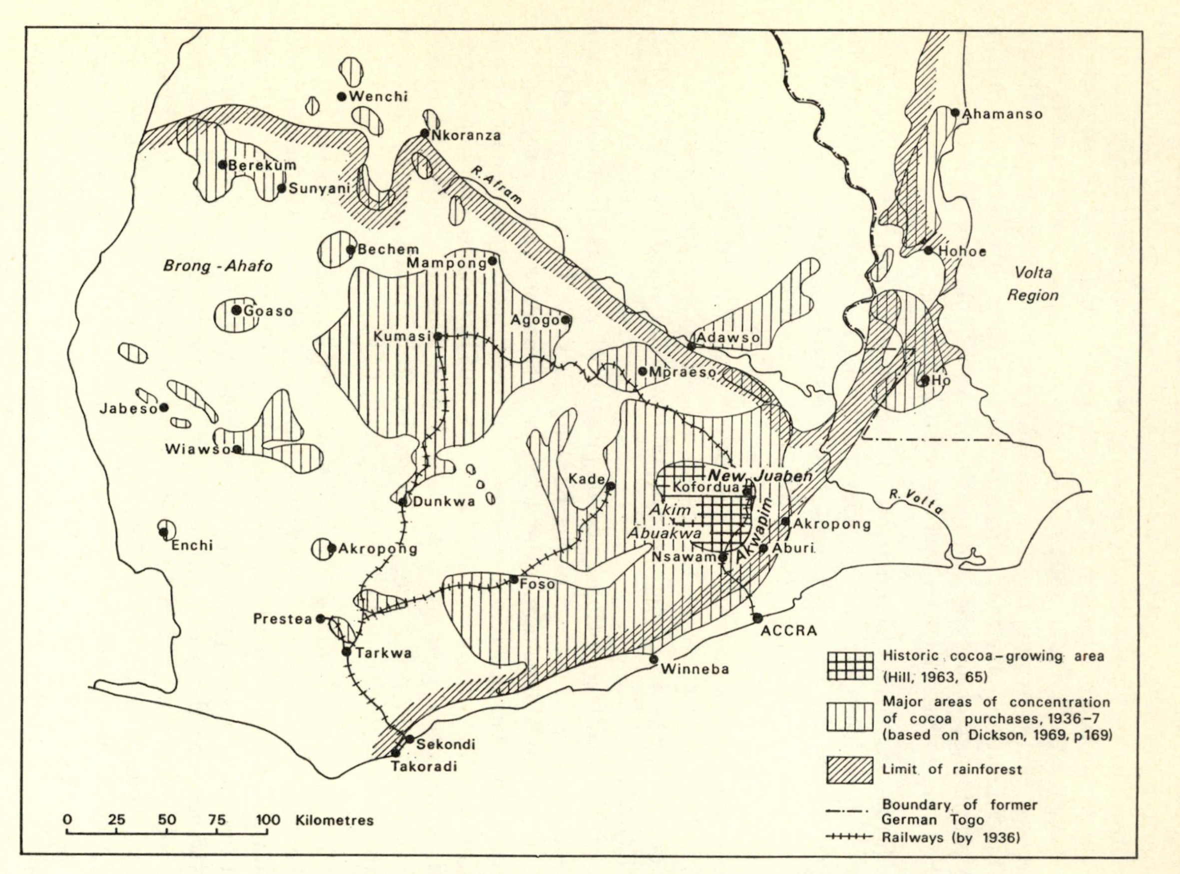

10.5 *Chief cocoa-purchase areas of Ghana, 1936–7.*

encouraged further expansion in the south-eastern plantings, the former encouraged advance west of Kumasi. Feeder roads extended the planting area even beyond the ecologically suitable rainforest 'zone' to 'islands' of rainforest in the savanna around Awawso. A number of cocoa areas was thus created in Ghana, each associated with the heavier soils of the basins (Morgan and Pugh 1969, 626) and with local commercial foci and feeder-road systems.

As planting advanced so production in the older areas declined, due partly to the increasing age of the trees and partly to disease and insect damage. Capsid attack was first recorded in 1907–8 and swollen shoot virus in 1936 (Wills 1962, 286–352) although the latter may have been established in the south-east from 1907. Swollen shoot spread rapidly after 1936 and was only limited by cutting down and burning several hundred thousand cocoa trees. Thousands of cocoa farmers had either to leave the districts affected or turn to food farming which by then, with the rapid growth of the towns, had become commercially attractive. Thus cocoa distribution did not exhibit as the distribution of some other crops, the persistence of a core area, but its decline and virtual abandonment.

Multiple enterprise distributions

Work on plotting and developing our understanding of multiple joint enterprise patterns has been extremely limited, partly because of the immense complexity of the task and partly because recognition of joint enterprise systems or farming types is debatable (see above, pp. 115–20).[1] The factors affecting the location of a given enterprise may differ considerably both in degree and kind from the factors affecting the location of another enterprise at the same point, even though both enterprises are parts of the same farm system. Differences in the degree of importance attached to different enterprises in a system differ enormously in a single locality among farms apparently of the same 'type'. Farm management is as important a locational factor as any. In consequence distribution patterns appear to be highly variable even where 'types' have apparently been recognized. It is hardly surprising that in the virtual infinity of farm 'types' that is statistically conceivable from the possibilities of enterprise combination alone, most classifications have depended on the notion of dominance of a single enterprise within each system, and in consequence, where mapping has taken place, have tended to show single enterprise distributions but limited to locations where those enterprises occupied the largest share of the farm

[1] Gregor's multiple-feature regions differ, being landscape regions including land capability, field and farm systems and general functions (Gregor 1970, 113–20).

area or business. Thus a map of cotton farming will show a more limited distribution than a map of the distribution of cotton which includes cotton grown on farms classified as belonging to other farm types.

The notion that there is a regular, purposeful integration of enterprises repeated on several farms in a locality may have some meaning here and there. It has been too easily accepted by many geographers and agriculturalists without adequate testing. It has had, especially in the past, some meaning in terms of rotations and use of labour and land, but in more recent years the ecological and organizational ties of farm enterprises have weakened and there are farms today with two or more enterprises produced by systems whose linkages are limited almost entirely to finance and use of farmer's time. The further notion that several farm types may be integrated to form 'the whole complex or system of agriculture of the wider region, e.g. store cattle and sheep grazing (one type of farming) in hill country, and stock fattening with cropping (another type) on lower ground, the two being integrated in an agricultural system' (Symons 1967, 95) appears to the authors of doubtful validity in its wider regional context when the kinds of linkage, e.g. between farms in the hills and farms in the lowlands, are so varied. Until much more work has been done on the nature of farming systems it is perhaps better to leave work on multiple enterprise distributions to the simple statistical recognition of certain broad areal coincidences (e.g. Coppock 1963, 202–4, and Weaver 1954*a* and *b*, 1956).

Trend surface analysis

For some geographers the notions of belts, areas of concentration, or regions are inadequate to provide a satisfactory descriptive framework for agricultural distributions. For them the transitions between one region and another, or rather the gradients of changing density within a distribution pattern are as important as any other aspect of study. It has been questioned whether regional boundaries do not obscure more than they indicate and claimed that very distinct boundaries are rare in agricultural distributions. Trend surface analysis has been suggested as an alternative (Tarrant 1969). The technique is to view a given enterprise distribution as a surface and then to calculate for that surface a best-fit plane or trend surface, in the manner in which a curve is fitted to a spread of dots in graphic analysis, except that in this case the fit is areal instead of being linear. The fitting of the plane is done by reducing to a minimum the sum of the squares of the vertical distances between the points indicating the distribution and the plane. Essentially a regression surface is calculated by using a polynomial function of the form (for one change of direction of slope):

$$X_n = c_0 + c_1U + c_2V + c_3U^2 + c_4UV + c_5UV^2$$

where X is any point, U and V refer to the orthogonal axes of the basic grid, and c_0, c_1, c_2, c_3, c_4, and c_5 are coefficients. For two or more changes of direction of slope the formula can be expanded (Chorley and Haggett 1965; King 1967; Cole and King 1968, 375–9). The method provides an indication of the general trends in any given distribution and assists the search for explanations. It also makes possible the precise delimitation of irregularities in the general pattern and their mapping as residuals. Again the location of distributional 'anomalies' may point the way to understanding the various factors governing a distribution. To some extent we may even obtain some idea of their order of importance by successive mappings of one, two, three directional surfaces, and so on, until a function is obtained which produces a surface nearly fitting the distribution. The technique has not yet been claimed as replacing that of mapping agricultural distributions or 'regions', but rather as complementary (Tarrant 1969).

Regional analysis

The techniques of regional analysis developed mainly by economists in order to obtain estimates of basic magnitudes in the spatial operation of an economy and for 'each region of a system' (Isard 1960, vii) have had only limited application so far in agricultural study. As these techniques are developed and find increasing use in national planning we are in danger of imposing a spatial order of our own on distributions for which we have created regional boundaries essentially of our own making. Thus we may claim for our economic regions a degree of homogeneity or nodality which may be more wished-for than real.

Nevertheless these techniques, more especially the use of linear programming, offer the practical possibility of viewing agriculture at a national level while taking some account of local variations. They should enable better forecasts of future productivity or of policy effects to be made by taking account of local variations in growth or response. Outstanding contributions have been made by Heady and Egbert (1964), and Egbert, Heady and Brokken (1964). In the latter study the attempt was made to assess interregional competition and the optimal space location of grain production in the United States. The work was prompted in part by the need to bring agricultural production in the United States more into line with food requirements. For the analysis 122 producing regions were used and 3 linear programming models with 500 restraints to specify which of the regions might most efficiently provide the national requirements of wheat, feed grains, cotton and soybeans in 1965. Henderson (1957 and 1959) tried to derive an optimal land-utilization pattern for the production of field crops by regions in the United States and thus for the United States as a whole. Snodgrass and French (1958) produced a study of interregional competition in dairying, using 24 regions, and seeking to discover minimal transport costs and the optimal interregional flow pattern (see also account of work by Henderson (1957), Snodgrass and French (1958) and Isard 1960, 481–8). Gregor (1963*a*) tried to define regional hierarchies of farm production and expenditure using physical regions, Morrill and Garrison (1960) have examined interregional trade patterns in wheat flour, Shaw (1970) the optimum pattern of potato production in Britain in relation to market transportation costs, while Schnittkar and Heady (1958) and Carter and Heady (1959) have applied input–output analysis to regional models and sectors of agriculture. For a thorough critical discussion of the value of regional input–output analysis and of regional linear programming, see Isard (1960, 309–74 and 413–92). More recently Isard *et al.* (1969) have attempted to develop a general theory of regional analysis based on a behavioural approach to the spatial pattern of decision-making authority and organization. Although primarily formulated in terms of urban decision-making situations, the theory could well provide a valuable conceptual framework for the explanation and formulation of regional agricultural policies and the response to these of individual producers. Gravity models have had hardly any application in regional agricultural geography, yet in view of the pioneer work of von Thünen on the effect of nodality, the constant interest since and the increasing attention on agricultural markets and urban–rural relationships, such application must come soon. Moreover, they have the advantage for many geographers that they represent behavioural rather than optimizing theory (Wilson 1967, quoted in Richardson 1969, 98).

Bibliography

AGRAWAL, R. C. and HEADY, E. O. (1968) Applications of game theory models in agriculture. *J. of Agric. Econ.* 19, 207–18.

AITCHISON, J. W. (1970) *The Farming Systems of Wales: a Study of Spatial Variability*. Publ. of the Dept. of Geogr., Univ. Coll. of Wales, Aberystwyth.

ALLAN, W. (1965) *The African Husbandman*. Edinburgh.

ASHTON, J. and CRACKNELL, B. E. (1960–1) Agricultural holdings and farm business structure in England and Wales. *J. of Agric. Econ.* 14, 472–506.

BACHMAN, L. L. and CHRISTENSEN, R. P. (1967) The economics of farm size, in *Agricultural Development and Economic Growth*, SOUTHWORTH, H. M. and JOHNSTON, B. F. (eds). Ithaca, 234–57.

BAINES, A. J. H. (1968) *A Century of Agricultural Statistics: Great Britain 1866–1966* (commentary). Ministry of Agriculture, Fisheries and Food, London.

BAKER, O. E. (1921) The increasing importance of the physical conditions in determining the utilization of land for agricultural and forest production in the United States. *Ann. Ass. Am. Geogr.* 11, 17–46.

BAKER, O. E. (1925) The potential supply of wheat. *Econ. Geogr.* 1, 15–52.

BAKER, O. E. (1926–32) Agricultural regions of North America. *Econ. Geogr.* 2, 460–93; 3, 50–86, 309–39, 445–65; 4, 44–73, 399–433; 5, 36–69; 6, 166–91, 278–309; 7, 109–53, 326–64; 8, 326–77.

BARKER, R. G. and WRIGHT, H. F. (1949) Psychological ecology and the problem of psychosocial development. *Child Dev.* 20, 131–43.

BARNES, F. A. and JEFFERY, D. M. (1964) Farming-type regions in the Dove Basin. *E. Midl. Geogr.* 3, 244–54.

BATEMAN, D. I. and WILLIAMS, H. T. (1966) Agriculture in the regional economy. *J. of Agric. Econ.* 17, 22–49.

BELSHAW, C. S. (1965) *Traditional Exchange and Modern Markets*. Englewood Cliffs, N.J.

BELSHAW, D. G. R. and JACKSON, B. G. (1966) Type-of-farm areas: the application of sampling methods. *Trans. Inst. Brit. Geogr.* 38, 89–93.

BENNETT, L. G. (1963) *The Diminished Competitive Power of the British Glasshouse Industry through Mal-location*. Univ. of Reading.

BEST, R. H. and GASSON, RUTH M. (1966) The changing location of intensive crops. *Stud. in Rur. Ld Use* 6, Wye College.

BIRCH, J. W. (1954) Observations on the delimitation of farming-type regions, with special reference to the Isle of Man. *Trans. Inst. Brit. Geogr.* 20, 141–58.

BIRCH, J. W. (1960) A note on the sample-farm survey and its use as a basis for generalized mapping. *Econ. Geogr.* 36, 254–9.

BLACK, J. D. (1953) *Introduction to Economics for Agriculture*. New York.

BLACK, G. J. (1965) Capital deployment on farms in theory and practice. I. *Fm Economist* 10, 475–84. II. *Fm Economist* 11, 1966, 10–24.

BLACK, M. (1968) Agricultural labour in an expanding economy. *J. of Agric. Econ.* 19, 59–76.

BLAIKIE, P. M. (1971) Spatial organization of agriculture in some north Indian villages: Part I. Trans. Inst. Brit. Geogr. 52, 1–40.

BLAUT, J. M. (1959) Microgeographic sampling: a quantitative approach to regional agricultural geography. *Econ. Geogr.* 35, 79–88.

BOARD, C. (1963) *Some methods of mapping farm-type areas*. Unpubl. pap. presented to the Fm Symp. of the Inst. Brit. Geogr., Aberystwyth.

BOSANQUET, C. I. C. (1968) Investment in agriculture. *J. of Agric. Econ.* 19, 3–12.

BOSERUP, E. (1965) *The Conditions of Agricultural Growth: the Economics of Agrarian Change under Population Pressure*. London.

BOWDEN, L. W. (1965) The diffusion of the decision to irrigate. *Dept. of Geogr., Univ. of Chicago, Res. Pap.*, 97.

BRINKMANN, T. (1922) *Die Oekonomik des Landwirtschaftlichen Betriebes*, part of Vol. VII of Grundriss der Sozialökonomik, Tubingen. Translation with introduction and notes by BENEDICT, E. T., STIPPLER, H. H., and BENEDICT, M. R., 1935.

BRITTON, D. K. and INGERSENT, K. (1964) Trends in concentration in British agriculture. *J. of Agric. Econ.* 16, 26–52.

BROOKFIELD, H. C. (1969) The environment as perceived. *Prog. in Geogr.* 1, 51–80.

BROWN, L. (1965) *Models for Spatial Diffusion Research*, Tech. Rep. No. 3, Spatial Diffusion Study, Dept. of Geogr., Northwestern Univ.

BUCHANAN, R. O. (1959) Some reflections on agricultural geography. *Geogr.* 44, 1–13.

BUCK, J. L. (1937) *Land Utilisation in China*, 3 vols, London.

BURBEE, H. (1965) Broiler production density, plant size, alternative operating plans and total unit costs. *North Carolina Exp. Stn Tech. Bull.*, No. 144.

BURTON, I. (1962) Types of agricultural occupance of flood plains in the United States. *Dept. of Geogr., Univ. of Chicago, Res. Pap.*, 75.

CAIRNCROSS, A. (1966) *Introduction to Economics*, 4th edn. London.

CANDOLLE, A. DE (1883) *Origine des Plantes Cultiveés*. Paris.

CARTER, H. A. and HEADY, E. O. (1959) An input–output analysis emphasizing regional and commodity sectors of agriculture. *Iowa Agric. and Home Econ. Exp. Stn Res. Bull.* 469, Ames, Iowa.

CATTELL, R. B. (1965) Factor analysis: an introduction to essentials. *Biom.* 21, 190–215.

CAVE, W. E. (1966) The impact of agricultural economics on a farmer. *J. of Agric. Econ.* 16, 470–5.

CHISHOLM, M. (1957) Regional variations in road transport costs: milk collection from farmers in England and Wales. *Fm Economist* 3, 30–8.

CHISHOLM, M. (1962) *Rural Settlement and Land Use: an Essay in Location*. London.

CHISHOLM, M. (1964) Problems in the classification and use of farming-type regions. *Trans. Inst. Brit. Geogr.* 35, 91–103.

CHISHOLM, M. (1966) *Geography and Economics*. London.

CHORLEY, R. J. and HAGGETT, P. (1965) Trend-surface mapping in geographical research. *Tran. Inst. Brit. Geogr.* 37, 47–67.

CHORLEY, R. J. and HAGGETT, P. (eds) (1967) *Models in Geography*. London.

CHRYST, W. E. (1965) Land values and agricultural income: a paradox? *J. of Fm Econ.* 47, 1265–77.

CHURCH, B. M. *et al.* (1968) A type of farming map based on agricultural census data. *Outlook on Agriculture* 5, 191–6.

CLARK, C. (1967) Von Thünen's 'Isolated State'. *Oxford Econ. Pap.* (New Series) 19, 370–7.

CLARK, C. (1969) The value of agricultural land. *J. of Agric. Econ.* 20, 1–23.

CLARK, C. and HASWELL, M. (1964) *The Economics of Subsistence Agriculture*. London (4th edition, 1970).

CLARK, P. J. and EVANS, F. C. (1954) Distance to nearest neighbour as a measure of spatial relationships in populations. *Ecol.* 35, 445–53.

CLAYTON, E. S. (1964) *Agrarian Development in Peasant Economies: Some Lessons from Kenya*. Oxford.

CLERY, P. A. and WOOD, P. D. P. (1965) Agricultural land values: 1962–1964. *J. of Agric. Econ.* 16, 567–72.

CLOUT, H. D. (1968) Planned and unplanned changes in French farm structures. *Geogr.* 53, 311–15.

COHEN, RUTH L. (1949) *The Economics of Agriculture*, (rev. ed.). Cambridge.

COLE, J. P. and KING, C. A. M. (1968) *Quantitative Geography*. London.

COLEMAN, ALICE (1961) The Second Land Use Survey: progress and prospect. *Geogr. J.* 127, 168–86.

CONKLIN, H. E. (1959) The Cornell System of land classification. *J. of Fm Economist* 41, 548–57.

COOLEY, W. W. and LOHNES, P. R. (1962) *Multivariate Procedures for the Behavioural Sciences*. New York.

COOPER, G. (1970) A glass roof over 50 acres. *Fmrs Wkly*, 27 March, xxi–xxvii.

COPPOCK, J. T. (1960) The parish as a geographical–statistical unit. *Tijdschr. Econ. Soc. Geogr.* 51, 317–26.

COPPOCK, J. T. (1962) Land use and land classification, in Classification of agricultural land in Britain. *Agric. Ld Serv. Tech. Rep.* 8, 65–80.

COPPOCK, J. T. (1964*a*) *An Agricultural Atlas of England and Wales*. London.

COPPOCK, J. T. (1964*b*) Crop-livestock and enterprise combinations in England and Wales. *Econ. Geogr.* 40, 65–81.

COPPOCK, J. T. (1964*c*) Post war studies in the geography of British agriculture. *Geogr. Rev.* 54, 409–26.

COPPOCK, J. T. (1965*a*) Regional differences in labour requirements in England and Wales. *Fm Economist* 10, 368–90.

COPPOCK, J. T. (1965*b*) The cartographic representation of British agricultural statistics. *Geogr.* 50, 101–14.

COPPOCK, J. T. (1968) The geography of agriculture. *J. of Agric. Econ.* 19, 153–75.

COPPOCK, J. T. (1971) *An Agricultural Geography of Great Britain*. London.

COURTENAY, P. P. (1965) *Plantation Agriculture*. London.

COWIE, W. J. G. and GILES, A. K. (1957) *Reasons for the Drift from the Land*. Univ. of Bristol.

COWLING, K. and METCALFE, D. (1968) Labour transfer from agriculture: a regional analysis. *Manchester Schl of Econ. and Soc. Stud.* 36, 27–48.

CURRY, L. (1962) The climatic resources of intensive grassland farming in the Waikato, New Zealand. *Geogr. Rev.* 52, 174–94.

CURRY, L. (1963) Regional variation in the seasonal programming of livestock farms in New Zealand. *Econ. Geogr.* 39, 95–118.

CYERT, R. M. and MARCH, J. G. (1963) *A Behavioural Theory of the Firm*. New York.

DALTON, G. E. (1967) The application of discounted cash flow techniques to agricultural investment problems. *J. of Agric. Econ.* 18, 363–74.

DAVIES, W. (1960) Pastoral systems in relation to world food supplies. *Adv. of Sci.* 17, 272–80.

DENMAN, D. R. (1965) Land ownership and the attraction of capital into agriculture: a British overview. *Ld Econ.* 41, 209–16.

DEXTER, K. and BARBER, D. (1967) *Farming for Profits*, 2nd edn. London.

DICKSON, K. B. (1969) *A Historical Geography of Ghana*. Cambridge.

DONALDSON, G. F. and WEBSTER, J. P. G. (1968) *An Operating Procedure for Simulation Farm Planning – Monte Carlo Method*. Wye College.

DONALDSON, J. G. S. and DONALDSON, F. (1969) *Farming in Britain Today*. London.

DORFMAN, R. (1968) Operations Research, chapter in *Surveys of Economic Theory, Resource Allocation*, Volume III. Am. Econ. Ass. and Roy. Econ. Soc., London, 29–74.

DUCKHAM, A. N. (1963) *Agricultural Synthesis: the Farming Year*. London.

DUCKHAM, A. N. (1967) Weather and farm management decisions, in *Weather and Agriculture*, TAYLOR, J. A. (ed.). London, 69–78.

DUCKHAM, A. N. and MASEFIELD, G. B. (1970) *Farming Systems of the World*. London.

DUMONT, R. (1957) *Types of Rural Economy: Studies in World Agriculture*. London.

DUNN, E. S. (1954) *The Location of Agricultural Production*. Gainesville.

EDWARDS, D. (1961) *An Economic Study of Small Farming in Jamaica*. Mona, Jamaica.

EGBERT, A. C., HEADY, E. O. and BROKKEN, R. F. (1964) Regional changes in grain production. *Iowa Agric. and Home Econ. Exp. Stn Res. Bull.* 521, Ames, Iowa.

ELLISON, W. (1953) *Marginal Land in Britain*. London.

EMERY, F. E. and OESER, O. A. (1958) *Information, Decision and Action: a Study of the Psychological Determinants of Changes in Farming Techniques*. Melbourne.

EMERY, F. E. and TRIST, E. L. (1965) The causal texture of organizational environments. *Human Relations* 18, 21–32. Reprinted in EMERY, F. E. (ed.), *Systems Thinking*. Harmondsworth, 1969, 241–57.

EPP, D. J. (1969) Some implications of the E.E.C.'s agricultural price policy. *Am. J. of Agric. Econ.* 51, 279–88.

ERICKSON, F. C. (1948) The broken Cotton Belt. *Econ. Geogr.* 24, 263–8.

EVANS, R. E. (1960) *Rations for Livestock*, Bull. No. 48, 15th ed. Ministry of Agriculture, Fisheries and Food, London.

FEDER, E. (1965) When is land reform a land reform? the Colombian case. *Am. J. Econ. and Sociol.* 24, 113–34.

FIELDING, G. J. (1964) The Los Angeles milkshed: a study of the political factor in agriculture. *Geogr. Rev.* 54, 1–12.

FIELDING, G. J. (1965) The role of government in New Zealand wheat growing. *Ann. Ass. Am. Geogr.* 55, 87–97.

FLORENCE, P. SARGANT (1944) The selection of industries suitable for dispersion into rural areas. *J. Roy. Stat. Soc.* 107, 93–107.

FLORENCE, P. SARGANT (1948) *Investment, Location and Size of Plant.* Cambridge.

FOUND, W. C. (1970) Towards a general theory relating distance between farm and home to agricultural production. *Geogr. Anal.* 2, 165–76.

FOX, R. W. (1967) Estimating the effects of the E.E.C. common grain policy. *J. of Fm Econ.* 49, 372–88.

GARDNER, T. W. (1957) A note on cereal prices and acreages. *J. of Agric. Econ.* 12, 361–8.

GARRISON, W. L. and MARBLE, D. F. (1957) The spatial structure of agricultural activities. *Ann. Ass. Am. Geogr.* 47, 137–44.

GASSON, RUTH (1966*a*) The changing location of intensive crops in England and Wales. *Geogr.* 61, 16–28.

GASSON, RUTH (1966*b*) Part-time farmers in south-east England. *Fm Economist* 11, 135–9.

GASSON, RUTH (1968) Occupations chosen by the sons of farmers. *J. of Agric. Econ.* 19, 317–26.

GASSON, RUTH (1969) Occupational immobility of small farmers. *J. of Agric. Econ.* 20, 279–88.

GEERTZ, C. (1963) *Agricultural Involution: The Process of Ecological Change in Indonesia.* Los Angeles.

GERLING, W. (1954) *Die Plantage.* Würzburg.

GLEAVE, M. B. and WHITE, H. P. (1969) Population density and agricultural systems in West Africa, in THOMAS, M. F. and WHITTINGTON, G. W. (eds), *Environment and Land Use in Africa.* London, 273–300.

GOULD, P. R. (1963) Man against his environment: a game theoretic framework. *Ann. Ass. Am. Geogr.* 53, 290–7.

GRAY, L. C. (1941) *History of Agriculture in the Southern United States to 1860*, Vol. II, New York. Reprint of the 1932 edition of the Carnegie Institution of Washington.

GREGOR, H. F. (1963*a*) Regional hierarchies in California's agricultural production; 1939–54. *Ann. Ass. Am. Geogr.* 53, 27–37.

GREGOR, H. F. (1963*b*) Industrialised dry-lot dairying: an overview. *Econ. Geogr.* 39, 299–318.

GREGOR, H. F. (1965) The changing plantation. *Ann. Ass. Am. Geogr.* 55, 221–38.

GREGOR, H. F. (1970) *Geography of Agriculture: Themes in Research.* Englewood Cliffs, N. J.

GRIFFIN, P. F., YOUNG, R. N. and CHATHAM, R. L. (1963) *Anglo-*

America: a Regional Geography of the United States and Canada. London.

GRIGG, D. (1963) Small and large farms in England and Wales: their size and distribution. *Geogr.* 48, 268–79.

GRIGG, D. (1966) Reply comment (to W. Bunge). *Ann. Ass. Am. Geogr.* 56, 376–77.

GRIGG, D. (1969) The agricultural regions of the world: review and reflections. *Econ. Geogr.* 45, 95–132.

GROTEWALD, A. (1959) Von Thünen in retrospect. *Econ. Geogr.* 35, 346–55.

GRUBER, J. and HEADY, E. O. (1968) Econometric analysis of the cattle cycle in the United States. *Iowa Agric. and Home Econ. Exp. Stn Res. Bull.* 564, Ames, Iowa.

GUITHER, H. D. (1963) Factors influencing farm operators' decisions to leave farming. *J. of Fm Econ.* 45, 567–76.

HÄGERSTRAND, T. (1952) The propagation of innovation waves. *Lund Studies in Geography,* Series B, Human Geography, 4, 3–19.

HÄGERSTRAND, T. (1953) *Innovationsförloppet ur Korologisk Synpunkt,* Lund. Translated as *Innovation Diffusion as a Spatial Process* by PRED, A. 1967, Chicago.

HAGGETT, P. (1965) *Locational Analysis in Human Geography.* London.

HALL, P. (ed.) (1966) *Von Thünen's Isolated State.* Oxford. Translated by Carla M. Wartenberg.

HARLE, J. T. (1968) Towards a more dynamic approach to farm planning. *J. of Agric. Econ.* 19, 339–46.

HARRIS, C. D. (1954) The market as a factor in the localization of industry in the United States. *Ann. Ass. Am. Geogr.* 44, 315–48.

HARRIS, D. R. (1969) The ecology of agricultural systems, in *Trends in Geography,* COOKE, R. U. and JOHNSON, J. H. (eds). London, 133–42.

HARTSHORNE, R. (1959) *Perspective on the Nature of Geography.* Chicago.

HARTSHORNE, R. and DICKEN, S. N. (1935) A classification of the agricultural regions of Europe and North America on a uniform statistical basis. *Ann. Ass. Am. Geogr.* 25, 99–120.

HARVEY, D. W. (1963) Locational change in the Kentish hop industry and the analysis of land use patterns. *Trans. Inst. Brit. Geogr.* 33, 123–44.

HARVEY, D. W. (1966) Theoretical concepts and the analysis of agricultural land use patterns in geography. *Ann. Ass. Am. Geogr.* 56, 361–74.

HARVEY, D. W. (1968) Pattern, process and the scale problem in geographical research. *Trans. Inst. of Brit. Geogr.* 45, 71–8.

HARVEY, D. W. (1969) *Explanation in Geography.* London.

HATHAWAY, D. E. (1963) *Government and Agriculture: Public Policy in a Democratic Society*. London.

HAYSTEAD, L. and FITE, G. C. (1955) *The Agricultural Regions of the United States*. Norman, Oklahoma.

HEADY, E. O. (1952) *Economics of Agricultural Production and Resource Use*. New York.

HEADY, E. O. (1954) Simplified presentation and logical aspects of linear programming techniques. *J. of Fm Econ.* 36, 1035–46.

HEADY, E. O. (1956) Relationship of scale analysis to productivity analysis, in *Resource Productivity, Returns to Scale, and Farm Size*, HEADY *et al.* (eds). Ames, Iowa.

HEADY, E. O. (1967) Trends and policies of agriculture in the U.S.A., in *Economic Change and Agriculture*, ASHTON, J. and ROGERS, S. J. (eds). Edinburgh, 209–28.

HEADY, E. O. and EGBERT, A. C. (1964) Regional programming of efficient agricultural production patterns. *Econometrica* 32, 374–86.

HENDERSON, J. M. (1957) The utilization of agricultural land: a regional approach. *Pap. Proc. Reg. Sci. Ass.* 3, 99–117.

HENDERSON, J. M. (1959) The utilization of agricultural land: a theoretical and empirical enquiry. *Rev. Econ. Stat.* 41, 242–60.

HENSHALL, JANET D. (1966) The demographic factor in the structure of agriculture in Barbados. *Trans. Inst. Brit. Geogr.* 38, 183–95.

HENSHALL, JANET D. (1967) Models of agricultural activity, in CHORLEY, R. J., and HAGGETT, P. (eds), *Models in Geography*. London, 425–58.

HENSHALL, JANET D. and KING, L. J. (1966) Some structural characteristics of peasant agriculture in Barbados, *Econ. Geogr.* 42, 74–84.

HILL, B. (1965) Supply responses in grain production in England and Wales, 1925–1963. *J. of Agric. Econ.* 16, 413–24.

HILL, P. (1963) *The Migrant Cocoa-Farmers of Southern Ghana: a Study in Rural Capitalism*. Cambridge.

HILTON, N. (1962) A new approach to agricultural land classification for planning purposes, in Classification of agricultural land in Britain. *Agric. Ld. Serv. Tech. Rep.* 8, 91–102.

HILTON, N. (1968) An approach to agricultural land classification in Great Britain. *Inst. Brit. Geogr. Spec. Publ.* No. 1, 127–44.

H.M.S.O. (1967) *Final Report: Statutory Smallholdings Provided by the Ministry of Agriculture, Fisheries and Food*, Cmnd. 3303.

HONKALA, K. (1969) Co-operation between farmers in Finland. *Sociologia Ruralis* 9, 235–51.

HOOVER, E. M. (1948) *The Location of Economic Activity*. New York.

HOUSTON, J. M. (1963) *A Social Geography of Europe*, reprinted with revisions. London.

HUNTER, J. M. (1963) Cocoa migration and patterns of land ownership in the Densu Valley near Suhum, Ghana. *Trans. Inst. Brit. Geogr.* 33, 61–87.

ISARD, W. (1960) *Methods of Regional Analysis: an Introduction to Regional Science*. New York.

ISARD, W., SMITH, T. E., ISARD, P., TUNG, T. H. and DACEY, M. (1969) *General Theory: Social, Political, Economic and Regional*. Cambridge, Massachusetts.

JACKS, G. V. (1946) *Land Classification for Land Use Planning*. Imperial Bureau of Soil Science, Harpenden.

JACKSON, B. G., BARNARD, C. and STURROCK, F. G. (1963) The pattern of farming in the eastern counties. *Occas. Pap., Fm Econ. Rev. Schl of Agric.*, Cambridge.

JACKSON, J. C. (1969) Towards an understanding of plantation agriculture. *Area* 4, 36–40.

JEVONS, W. S. (1871) *The Theory of Political Economy*. London.

JOHNSON, G. L., HALTER, A. N., JENSEN, M. R. and THOMAS, D. (1961) *A Study of Managerial Processes of Mid-Western Farmers*. Ames, Iowa.

JONASSON, O. (1925) Agricultural regions of Europe. *Econ. Geogr.* 1, 277–314.

JONES, G. E. (1963) The diffusion of agricultural innovations. *J. of Agric. Econ.* 15, 59–69.

JONES, G. E. (1965) In discussion in The geographical typology of agriculture, *Agricultural Geography*, 59–74, I. G. U. Symp. of 1964, Dept. of Geogr., Univ. of Liverpool, p. 64.

JONES, G. E. (1967) The adoption and diffusion of agricultural practices. *Wld Agric. Econ. and Rur. Sociol. Abs* 9, Review Article No. 6, 1–34.

JONES, R. BENNETT (1954) *The Pattern of Farming in the East Midlands*. Schl of Agric., Univ. of Nottingham.

JONES, R. BENNETT (1957) Farm classification in Britain – an appraisal. *J. of Agric. Econ.* 12, 201–224.

JONES, R. BENNETT (1969) Stability in farm incomes. *J. of Agric. Econ.* 20, 111–24.

KAY, G. (1969) Agricultural progress in Zambia, in THOMAS, M. F. and WHITTINGTON, G. W., *Environment and Land Use in Africa*. London, 495–524.

KENDALL, M. G. (1939) Geographical distribution of crop productivity in England. *J. Roy. Stat. Soc.* 102, 21–62.

KENDALL, M. G. (1957) *A Course in Multivariate Analysis*, London.

KING, C. A. M. (1967) An introduction to trend surface analysis. *Bull. Quant. Data for Geogr.* 12, Univ. of Nottingham.

KLAGES, E. (1949) *Ecological Crop Geography*. London.

KLAYMAN, M. I. (1960) *International Index Numbers of Food and*

Agricultural Production. F.A.O. Mon. Bull. Agric. Econ. and Stat. 9, Rome.

KOSTROWICKI, J. (1964*a*) Geographical typology of agriculture: principles and methods. *Geogr. Polonica* 2, 159–67.

KOSTROWICKI, J. (1964*b*) A geographical typology of agriculture: principles and methods (abstract). *Abstract of Papers* 20th I.G.U. Congress, London.

KOSTROWICKI, J. (1966) *Principles, Basic Notions and Criteria of Agricultural Typology*, Conclusions drawn from the answers to the Commission's Questionnaire No. 1, I.G.U., Commis. Agric. Typology, Warsaw.

KOSTROWICKI, J. (1969) Agricultural Typology. *Bull. I.G.U.* 20, 36–40.

KOSTROWICKI, J. (1970) Types of agriculture in Poland. A preliminary attempt at a typological classification. *Geogr. Polonica* 19, 99–110.

KRUEGER, R. (1959) The disappearing Niagara Fruit Belt. *Can. Geogr. J.* 58, 102–14.

KRUMBEIN, W. C. (1955) Statistical analysis of facies maps. *J. Geol.* 63, 452–70.

KRUMBEIN, W. C. (1956) Regional and local components in facies maps. *Bull. Am. Ass. Petrol. Geol.* 40, 2163–94.

LAND UTILISATION SURVEY (1941) *Types of Farming: England and Wales*, coloured map, 1:633,600. London.

LANGLEY, J. A. (1966) Risk, uncertainty and the instability of incomes in agriculture. *Occas. Pap.* 2, Dept. of Agric. Econ., Univ. of Exeter.

LANGLEY, J. A. and LUXTON, H. W. B. (1958) *The Farms of Dorset*, Report No. 104, Dept. of Econ. (Agric. Econ.), Univ. of Bristol.

LAUT, P. (1968) *Agricultural Geography* Vol. I Systems, subsistence and plantation agriculture. Vol. II. Mid-latitude commercial agriculture. Melbourne.

LEWIS, A. B. (1969) Economic land classification in the Far East with references to basic studies elsewhere. *Wld Agric. Econ. and Rur. Sociol. Abs* Review Article No. 9, 11, 1–22.

LONG, M. F. (1968) Why peasant farmers borrow. *Am. J. Agric. Econ.* 50, 991–1008.

LONG, W. M. (1957) The problem of the small farm. *J. Roy. Agric. Soc.* 118, 22–9.

LÖSCH, A. (1954) *The Economics of Location*. New Haven.

MACKNEY, D. and BURNHAM, C. P. (1964) *The Soils of the Church Stretton District of Shropshire*. Memoir of the Soil Survey of England and Wales, Harpenden.

MAJUMDAR, D. (1965) Size of farm and productivity: a problem of Indian peasant agriculture. *Economica* 32, 161–73.

MALTHUS, T. (1798) *Principles of Population*. London.

MARSHALL, A. (1890) *The Principles of Economics*. London.

MCCARTY, H. H. and LINDBERG, J. B. (1966) *A Preface to Economic Geography*. Englewood Cliffs.

MCCLEMENTS, L. D. (1969) The specification of pig supply models. *Fm Economist* 11, 425–7.

MACFADYEN, A. (1964) Energy flow in ecosystems and its exploitation by grazing, in *Grazing in Marine and Terrestial Environments*, CRISP, D. J. (ed), Brit. Ecol. Soc. Symp. 4. Oxford, 3–20.

MERCER, W. B. (1963) *A Survey of the Agriculture of Cheshire*, Roy. Agric. Soc. England, County Agric. Surv., No. 4. London.

METCALF, D. (1969) *The Economics of Agriculture*. Harmondsworth.

MEYER, J. R. (1963) Regional economics: a survey. *Am. Econ. Rev.* 53, 19–54.

MILTHORPE, F. L. (1965) Crop responses in relation to the forecasting of yields, in JOHNSON, C. G. and SMITH, L. P. (eds), *The Biological Significance of Climatic Changes in Britain*, 119–28.

MINISTRY OF AGRICULTURE, FISHERIES AND FOOD (1964) *Hill Farming Act*. London.

MINISTRY OF AGRICULTURE, FISHERIES AND FOOD (1968*a*) Grass and Grassland, Bulletin 154, 4th edn, 1966, reprinted with minor corrections. London.

MINISTRY OF AGRICULTURE, FISHERIES AND FOOD (1968*b*) *Agricultural Land Classification Map of England and Wales, Explanatory Note*, Agric. Ld Serv. London.

MINISTRY OF AGRICULTURE, FISHERIES AND FOOD (1969) *Types of Farming Maps of England and Wales*. London.

MIRACLE, M. P. (1968) Subsistence agriculture, analytical problems and alternative concepts. *Amer. J. of Agric. Econ.* 50, 292–310.

MOERMAN, M. (1968) *Agricultural Change and Peasant Choice*. Berkeley.

MOORE, L. (1960) The effect of changes in crop acreage on yield. *Fm Economist* 4, 383–5.

MORGAN, W. B. (1955) Farming practice, settlement patterns and population density in south-eastern Nigeria. *Geogr. J.* 121, 320–33.

MORGAN, W. B. (1969) The zoning of land use around rural settlements in Tropical Africa, in THOMAS, M. F. and WHITTINGTON, G. W. (eds), *Environment and Land Use in Africa*. London, 301–19.

MORGAN, W. B. and PUGH, J. C. (1969) *West Africa*. London.

MORRILL, R. L. and GARRISON, W. L. (1960) Projections of interregional patterns of trade in wheat and flour. *Econ. Geogr.* 36, 116–26.

MOSS, R. P. and MORGAN, W. B. (1970) Soils, plants and farmers in West Africa, in GARLICK, J. P. (ed), *Human Ecology in the Tropics*. Oxford, 1–31.

MUNTON, R. J. C. and NORRIS, J. M. (1969) The analysis of farm organisation: an approach to the classification of agricultural land in Britain. *Geografiska. Ann.* 52B, 95–103.

NALSON, J. S. (1968) *Mobility of Farm Families.* Manchester.

NAPOLITAN, L. and BROWN, C. J. (1963) A type of farming classification of agricultural holdings in England and Wales according to enterprise patterns. *J. of Agric. Econ.* 15, 595–616.

NIX, J. (1969) *Farm Management Pocketbook,* 3rd edn (2nd edn, 1968). Wye College.

O'DWYER, T. (1968) A comparison of costs of milk assembly in a cooperative society area. *Irish J. of Agric. Econ. and Rur. Soc.* 1, 221–43.

O'DWYER, T. (1970) The market for agricultural produce. *Area* 1, 52–3.

O.E.C.D. (1965) *Agriculture and Economic Growth.* Paris.

OJALA, E. M. (1967) Some current issues of international commodity policy. *J. of Agric. Econ.* 18, 27–46.

OLMSTEAD, C. W. and MANLEY, V. P. (1965) The geography of input, output and scale of operation in American agriculture. *Agricultural Geography,* 37–47, I.G.U. Symp. of 1964, Dept. of Geogr., Univ. of Liverpool.

OOI JIN BEE (1959) *Land, People and Economy in Malaya.* London.

PAHL, R. E. (1965) *Urbs in Rure,* Geogr. Pap. No. 2, L.S.E.

PATRICK, G. F., and EISGRUBER, L. A. (1969) The impact of managerial ability and capital structure on growth of the farm firm. *Am. J. Agric. Econ.* 51, 491–505.

PEDLER, F. J. (1955) *Economic Geography of West Africa.* London.

PERLOFF, H. S., DUNN, E. S. JR, LAMPARD, E. E. and MUTH, R. F. (1960) *Regions, Resources and Economic Growth.* Baltimore.

PRED, A. (1967) Behaviour and location: foundations for a geographic and dynamic location theory, Part I. *Lund Stud. Geogr.* Ser. B, 27.

PRED, A. (1969) Behaviour and location: foundations for a geographic and dynamic location theory, Part II. *Lund Stud. Geogr.* Ser. B, 28.

PRUNTY, M., JR, (1951) Recent quantitative changes in the cotton regions of the United States. *Econ. Geogr.* 27, 189–208.

PUTNAM, G. E. (1923) *Supplying Britain's Meat.* London.

REES, G. and WISEMAN, J. (1969) London's commodity markets. *Lloyds Bk Rev.* 91, 22–45.

RENBORG, U. (1962) *Studies of the Planning Environment of the Agricultural Firm.* Uppsala.

RICARDO, D. (1817) *On the Principles of Political Economy and Taxation.* London. Edited by P. Sraffa as Vol. I, *The Works and Correspondence of David Ricardo.* Cambridge, 1951.

RICHARDSON, H. W. (1969) *Elements of Regional Economics.* Harmondsworth.

ROGERS, E. M. (1958) Categorizing the adoption of agricultural practices. *Rur. Sociol.* 23, 345–54.

ROGERS, E. M. (1962) *Diffusion of Innovations*. Glencoe, Minnesota.

ROXBY, P. M. (1925–6) The theory of natural regions. *Geogr. Teacher* 13, 376–82.

ROYEN, W. VAN and BENGTSON, N. A. (1964) *Fundamentals of Economic Geography*, 5th edn. Englewood Cliffs, N.J.

RUTHERFORD, D. J. (1966) The double-cropping of wet padi in Penang, Malaya. *Geogr. Rev.* 56, 239–55.

RUMMEL, R. J. (1968) Understanding factor analysis. *J. of Conflict Resolution* 11, 444–80.

RYAN, B. and GROSS, N. C. (1943) The diffusion of hybrid seed corn in two Iowa communities. *Rur. Sociol.* 8, 15–24.

SAARINEN, T. F. (1966) Perception of the drought hazard on the Great Plains. *Res. Pap.* 106, Dept. of Geogr., Univ. of Chicago.

SAUER, C. O. (1952) Agricultural origins and dispersals. *American Geographical Society, Bowman Memorial Lectures*, No. 2.

SCHNITTKAR, J. A. and HEADY, E. O. (1958) Applications of input-output analysis to a regional model stressing agriculture. *Iowa Agric. Home Econ. Exp. Stn Res. Bull.* 454. Ames, Iowa.

SCHULTZ, T. W. (1964) *Transforming Traditional Agriculture*. New Haven.

SCOLA, P. M. (1952) Problems of farm classification. *J. of Agric. Econ.* 10, 4–16.

SCOLA, P. M. and MACKENZIE, A. M. (1952) *Types of Farming in Scotland*. Edinburgh.

SEWELL, W. R. D. (ed) (1966) Human dimensions of weather modification, *Res. Pap.*, 105, Dept. of Geogr. Univ. of Chicago.

SHAW, J. R. (1970) The location of maincrop potato production in Britain – an application of linear programming. *J. of Agric. Econ.* 21, 267–82.

SHAW, L. M. (1964) The effects of weather on agricultural output: a look at methodology. *J. of Fm Econ.* 46, 218–30.

SHEPHERD, G. S. (1964) *Farm Policy: New Directions*. Iowa.

SIMMONS, I. (1966) Ecology and land use. *Trans. Inst. Brit. Geogr.* 38, 59–72.

SIMON, H. A. (1968) Theories of decision making in economics and behavioural science, in *Surveys in Economic Theory, Resource Allocation*, Vol. III, Am. Econ. Ass. and Roy. Econ. Soc. London, 1–28.

SIMPSON, E. S. (1957) The Cheshire grass-dairying region. *Trans. Inst. of Brit. Geogr.* 23, 141–62.

SIMPSON, E. S. (1959) Milk production in England and Wales: a study in collective marketing. *Geogr. Rev.* 95–111.

SIMPSON, E. S. (1965) In discussion in The geographical typology of agriculture, *Agricultural Geography* 59–74, I.G.U. Symposium of 1964, Dept. of Geogr., Univ. of Liverpool, p. 66.

SINCLAIR, R. (1967) Von Thünen and urban sprawl. *Ann. Ass. Am. Geogr.* 47, 72–87.

SLICHER VAN BATH, B. H. (1963) *The Agrarian History of Western Europe, A.D. 500–1850*, London. Translated by Olive Ordish from the Dutch *De Agrarische Geschiedenis van West-Europa, 500–1850.*

SMITH, J. RUSSELL (1925) *North America.* New York.

SMITH, L. P. F. (1961) *The Evolution of Agricultural Cooperation.* Oxford.

SMITH, W. (1955) The location of industry. *Trans. Inst. Brit. Geogr.* 21, 1–18.

SNODGRASS, M. M. and FRENCH, C. E. (1958) *Linear Programming Approach to Interregional Competition in Dairying.* Purdue Univ., Agric. Exp. Stn Lafayette, Ind.

SPENCER, J. E. and HORVARTH, R. J. (1963) How does an agricultural region originate? *Ann. Ass. Am. Geogr.* 53, 74–92.

STAMP, L. D. (1943) *Kent*, County Report No. 85 of the Land Utilisation Survey of Britain. London.

STAMP, L. D. (1946) *The Land of Britain and How It Is Used*, Published for the British Council. London.

STAMP, L. D. (1948) *The Land of Britain: its Use and Misuse*, 3rd Edn, 1962. London.

STAMP, L. D. (1958) The measurement of land resources. *Geogr. Rev.* 48, 1–15.

STAMP, L. D. (1960) *Applied Geography.* Harmondsworth.

STENNING, D. J. (1959) *Savannah Nomads.* London.

STEWARD, J. H. (1960) Perspectives on plantations. *Revta Geogr.* No. 52, 26, 77–85.

STEWART, I. M. T. (1964) Farmers' transport costs in the highlands. *Scot. Agric. Econ.* 14, 223–45.

STURROCK, F. G. (1965) The optimum size of the family farm. *Occas. Pap.* No. 9, Schl of Agric., Fm Econ. Bran., Univ. of Cambridge.

SYKES, G. (1963) *Poultry: a Modern Agribusiness.* London.

SYMONS, L. (1967) *Agricultural Geography.* London.

TAKES, CH. A. P. (1964) Problems of rural development in southern Nigeria. *Tijdschr. K. Ned. Aardrijsk. Genoot.* 81, 438–52.

TARRANT, J. R. (1969) Some spatial variations in Irish agriculture. *Tijdschr. Econ. Soc. Geogr.* 60, 228–37.

TAYLOR, J. A. (1952) The relation of crop distributions to the drift pattern in south-west Lancashire. *Trans. Inst. Brit. Geogr.* 18, 77–91.

TAYLOR, J. A. (1969) *Weather and Agriculture.* London.

TEFERTILLER, R. R. and HILDRETH, R. J. (1962) Importance of weather variability on management decisions. *J. of Fm Econ.* 43, 1163–71.

THOMAS, D. (1963) *Agriculture in Wales during the Napoleonic Wars: a Study in the Geographical Interpretation of Historical Sources.* Cardiff.

THORNS, D. C. (1968) The influence of social change upon the farmer. *Fm Economist* 11, 337–44.

THORNTON, D. S. (1962) The study of decision making and its relevance to the study of farm management. *Fm Economist* 10, 40–56.

THÜNEN, J. H. VON (1826) *Der Isolierte Staat in Beziehung auf Landwirtschaft und Nationalökonomie,* Pt I, Rostock. Collected Edition, Pts I, II, and III, 1876. Berlin.

UPTON, M. (1966) Tree crops: a long term investment. *J. of Agric. Econ.* 17, 82–90.

UPTON, M. (1967) *Agriculture in South-Western Nigeria,* Development Studies No. 3, Dept. of Agric. Econ., Univ. of Reading.

URQUHART, G. A. (1965) The amalgamation of agricultural holdings: a study of two parishes in western Aberdeenshire. *J. of Agric. Econ.* 16, 405–12.

VAVILOV, N. I. (1949–50) Origin, variation, immunity and breeding of cultivated plants. *Chron. Bot.* 13.

WATSON, J. A. S. and MORE, J. A. (1962) *Agriculture: The Science and Practice of British Farming,* 11th edn. Edinburgh.

WEAVER, J. C. (1954*a*) Crop-combination regions in the Middle West. *Geogr. Rev.* 44, 175–200.

WEAVER, J. C. (1954*b*) Crop-combination regions for 1919 and 1929 in the Middle West. *Geogr. Rev.* 44, 560–72.

WEAVER, J. C. *et al.* (1956) Livestock units and combination regions in the Middle West. *Econ. Geogr.* 32, 237–59.

WEBER, A. (1909) *Ueber der Standort der Industrien, Pt. I, Reine Theorie des Standorts,* Tubingen. Translated by C. J. Freidrich as *Alfred Weber's Theory of the Location of Industries,* Chicago, 1929.

WEEKS, W. G. R. and BRAYSHAW, G. H. (1966) *Fat Cattle Auction Markets in Great Britain.* Newcastle-upon-Tyne.

WHETHAM, E. H. (1966) Diminishing returns and agriculture in northern Nigeria. *J. of Agric. Econ.* 17, 151–8.

WHITBY, M. C. (1968) Lessons from Swedish farm structure policy. *J. of Agric. Econ.* 19, 279–99.

WHITE, G. L., FOSCUE, E. J. and MCKNIGHT, T. L. (1964) *Regional Geography of Anglo-America,* 3rd edn. London.

WHITTLESEY, D. (1936) Major agricultural regions of the earth. *Ann. Ass. Am. Geogr.* 26, 199–240.

WHITTLESEY, N. K. (1967) Cost and efficiency of alternative land retirement programs. *J. of Fm Econ.* 49, 351–9.

WIBBERLEY, G. P. (1954) Some aspects of problem rural areas in Britain. *Geogr. J.* 120, 43–61.

WILKINSON, A. E. (1969) Operations and costs of egg packing stations. *J. of Agric. Econ.* 20, 331–44.

WILLIAMS, C. M. and HARDAKER, J. B. (1966) The small Fen farm. *Occas. Pap.* 10, Schl of Agric., Fm Econ. Bran., Univ. of Cambridge.

WILLS, J. B. (ed.) (1962) *Agriculture and Land Use in Ghana.* London.

WILSON, A. J. (1967) *Towards Comprehensive Planning Models.* Conference of the Regional Science Association (mimeo).

WOLPERT, J. (1964) The decision process in a spatial context. *Ann. Ass. Am. Geogr.* 54, 537–58.

YUDELMAN, M. (1964) *Africans on the Land.* Cambridge, Massachusetts.

ZVORYKIN, K. V. (1963) Scientific principles for an agro-production classification of lands. *Soviet Geogr.* 4, 3–9.

Appendix

The metric system: conversion factors and symbols

In common with several other text book series *The Field of Geography* uses the metric units of measurement recommended for scientific journals by the Royal Society Conference of Editors.* For geography texts the most commonly used of these units are:

Physical quantity	*Name of unit*	*Symbol for unit*	*Definition of non-basic units*
length	metre	m	basic
area	square metre	m^2	basic
	hectare	ha	10^4 m^2
mass	kilogramme	kg	basic
	tonne	t	10^3 kg
volume	cubic metre	m^3	basic-derived
	litre	l	10^{-3} m^3, 1 dm^3
time	second	s	basic
force	newton	N	kg m s^{-2}
pressure	bar	bar	10^5 Nm^{-2}
energy	joule	J	kgm^2 s^{-2}
power	watt	W	kgm^2 s^{-3} = Js^{-1}
thermodynamic temperature	degree Kelvin	°K	basic
customary temperature, t	degree Celsius	°C	t /°C = T /°K − 273·15

* Royal Society Conference of Editors, *Metrication in Scientific Journals*, London, 1968.

Fractions and multiples

Fraction	*Prefix*	*Symbol*	*Multiple*	*Prefix*	*Symbol*
10^{-1}	deci	d	10	deka	da
10^{-2}	denti	c	10^{2}	hecto	h
10^{-3}	milli	m	10^{3}	kilo	k
10^{-6}	micro	μ	10^{6}	mega	M

The gramme (g) is used in association with numerical prefixes to avoid such absurdities as mkg for μg or kkg for Mg.

Conversion of common British units to metric units

Length

1 mile = 1·609 km
1 furlong = 0·201 km
1 chain = 20·117 m
1 fathom = 1·829 m
1 yard = 0·914 m
1 foot = 0·305 m
1 inch = 25·4 mm

Area

1 sq mile = 2·590 km²
1 acre = 0·405 ha
1 sq foot = 0·093 m²
1 sq inch = 645·16 mm²

Mass

1 ton = 1·016 t
1 cwt = 50·802 kg
1 stone = 6·350 kg
1 lb = 0·454 kg
1 oz = 28·350 g

Mass per unit length and per unit area

1 ton/mile = 0·631 t/km
1 lb/ft = 1·488 kg/m
1 ton/sq mile = 392·298 kg/km²
1 cwt/acre = 125·535 kg/ha

Volume and capacity

1 cubic foot = 0·028 m³
1 cubic inch = 1,638·71 mm³
1 US barrel = 0·159 m³
1 bushel = 0·036 m³
1 gallon = 4·546 l
1 US gallon = 3·785 l
1 quart = 1·137 l
1 pint = 0·568 l
1 gill = 0·142 l

Velocity

1 m.p.h. = 1·609 km/h
1 ft/s = 0·305 m/s
1 UK knot = 1·853 km/h

Mass carried × distance

1 ton mile = 1·635 t km

Force

1 ton-force	= 9·964 kN	1 poundal	= 0·138 N
1 lb-force	= 4·448 N	1 dyn	= 10^{-5} N

Pressure

1 ton-force/ft^2	= 107·252 kN/m^2	1 pdl/ft^2	= 1·488 N/m^2
1 lb-force/in^2	= 68·948 mbar		

Energy and power

1 therm	= 105·506 MJ	1 Btu	= 1·055 kJ
1 hp	= 745·700 W(J/s)	1 ft lb-force	= 1·356 J
	= 0·746 kW	1 ft pdl	= 0·042 J
1 hp/hour	= 2·685 MJ	1 cal	= 4·187 J
1 kWh	= 3·6 MJ	1 erg	= 10^{-7} J

Metric units have been used in the text wherever possible. British or other standard equivalents have been added in brackets in a few cases where metric units are still only used infrequently by English-speaking readers.

Index